# Out of Scope: Become the Strategic PM the Business Needs

By

## Lynn R. Squire

# Disclaimer

# Table of Contents

# Prologue

Project management was never meant to be limited to tracking tasks and updating timelines. Yet, in many organizations, project managers have been reduced to little more than coordinators with tools, expected to produce status reports and keep meetings running on schedule. This perception has cost both companies and PMs more than they realize. While projects may be delivered "on time," they often fall short of achieving meaningful business outcomes. Strategic opportunities are often missed because PMs are excluded from key conversations. As a result, talented professionals with the potential to drive real business impact remain confined to operational roles.

The business landscape has changed. Organizations now operate in an environment where market shifts are rapid, customer expectations evolve daily, and innovation cycles are relentless. Executives are no longer impressed by Gantt charts or detailed progress updates. They are asking tougher questions: Will this project move the needle on revenue or customer growth? Is it worth the investment? What happens if we don't do it? In this environment, a project manager who can confidently answer these questions and align project work with business strategy is no longer optional—they're a necessity.

This shift is already underway. Across industries, the most respected project managers are stepping into roles that look more like business strategists than task managers. They influence priorities, challenge assumptions, and assist leadership in making decisions about where to invest resources. They understand the company's business model, analyze a profit-and-loss statement, and know how to translate technical milestones into metrics that executives care about. These are the PMs who get invited to strategic meetings, who are trusted with high-visibility initiatives, and who become indispensable to leadership teams.

But this evolution does not happen automatically. It requires a deliberate shift in mindset and skillset. It entails moving beyond tools and frameworks to develop business acumen, financial literacy, and the ability to communicate in the language of executives. This means learning to lead with influence, not authority, and to make decisions when data is incomplete and trade offs are inevitable.

This book is designed to help you make that shift. It will show you how to move from managing tasks to driving outcomes, from reporting on progress to shaping business results. Through practical strategies, real-world scenarios, and tools you can apply, you will learn how to position yourself as a true strategic partner to your organization.

The era of the task-focused project manager is ending. The future belongs to those who can think beyond the Gantt chart, align their work with business priorities, and lead with strategic insight. If you are ready to become that kind of project leader, this book will guide you there.

# Part I

# The Problem: Why PMs Are Being Left Out of Strategy

# Chapter 1 – From Task Master to Strategic Asset: The Identity Gap

The role of project managers has evolved, but perceptions lag. In today's volatile environment, delivering on time and on budget isn't enough—leadership expects projects to drive revenue, improve customer experience, or strengthen competitive advantage.

Yet, many project managers are still viewed as coordinators rather than partners in achieving strategic goals. Research shows that in many companies, PMs are excluded from key decision-making because they are seen as operational rather than strategic[1]. At the same time, automation has taken over much of the traditional tracking and reporting work, raising a critical question: what unique value does a project manager add if they cannot influence outcomes?

This identity gap between organizational need and perceived role is the largest barrier to credibility at leadership levels. Closing it requires understanding how the discipline has transformed and why old perceptions persist.

## The Evolution vs. The Perception Gap

For decades, PMs were judged by efficiency: schedules met, budgets contained, resources coordinated. In stable industries, this narrow, operational definition worked.

---

[1] Taylor, Paul. "When Leaders Get Stuck in the Weeds." *LinkedIn*, June 28, 2017. https://www.linkedin.com/pulse/when-leaders-get-stuck-weeds-paul-taylor/

But today's environment demands more. Markets shift quickly, customer expectations outpace product cycles, and organizations need measurable value, not just completed projects. Modern PMs are expected to act as "mini-CEOs" of initiatives—aligning delivery with strategy, prioritizing outcomes, and demonstrating business impact.

Despite this shift, many organizations still see PMs through an outdated lens: administrators of delivery rather than strategic assets. Until this perception changes, the full potential of project management will remain untapped.

## The Perception Lag

Despite growing evidence of the value strategic PMs can bring, many executives and even PMOs still view PMs primarily as task managers. A recurring sentiment from the discussion forums is that PMs are "replaceable coordinators"—easily substituted by anyone familiar with project tracking software[2]. This perception persists partly because PMs often communicate in operational terms: status updates, percentage completion, and risk logs. For executives, who are focused on business growth, these updates rarely answer the most critical question: Why does this project matter to the organization?

Experts argue that leaders often contribute to this perception problem by dragging PMs deeper into operational details[3]. Instead of involving them in conversations about trade-offs or

---

[2] "Why Project Managers Are Not Respected ?" 2017.
ProjectManagement.com. September 2017.
https://www.projectmanagement.com/discussion-topic/69961/why-project-managers-are-not-respected--# .

[3] Taylor, "Stuck in the Weeds"

strategic direction, they use them as conduits for information flow. This dynamic reinforces the idea that PMs are administrators rather than strategic partners.

Technology has also played a paradoxical role. Tools like Jira, Asana, and Monday.com have made tracking and reporting almost fully automated. Experts have warned that "the role of the project manager is disappearing" for those who don't evolve[4].

*If software can generate schedules and update task progress instantly, what unique value does a PM add?* The answer lies in business alignment and strategic influence—capabilities that no software can replicate.

## A Shifting Expectation

Executives in forward-thinking organizations are starting to look for more than progress reports. They want project managers who can anticipate market shifts, connect deliverables to revenue and customer impact, and provide insight on resource allocation.

However, research highlights that many PMs hesitate to step into this strategic space because they are still evaluated primarily on operational efficiency[5]. Until PMs change how they position themselves, this perception gap will persist.

---

[4] "LinkedIn." 2025. Linkedin.com. 2025.
https://www.linkedin.com/posts/ddrofa_the-role-of-the-project-manager-is-disappearing-activity-7316029454676328448-ryqK/.

[5] Maylor, Harvey. "Beyond the Gantt Chart: Project Management Moving On." European Management Journal 19, no. 1 (2001): 92–100. https://www.scribd.com/document/6596337/Beyond-the-Gantt-Chart-Project-Management-Moving-On?v=0.148.

## Old PM Role vs. Strategic PM Role

| Traditional PM Role | Strategic PM Role |
|---|---|
| Focused on deadlines and budgets | Focused on business outcomes and value delivery |
| Reports project progress to stakeholders | Advises leadership on strategic tradeoffs and priorities |
| Communicates in tasks, risks, and timelines | Communicates in ROI, customer impact, and strategic benefits |
| Administrative coordinator | Trusted business partner |

Closing this identity gap is the first step in transitioning from being seen as a task master to becoming a strategic asset. PMs who understand this shift and can communicate in the language of strategy are the ones executives turn to when decisions truly matter.

## Why PMs Are Stuck as 'Human Calendars'

Projects today demand more than just on-time delivery. They require outcomes tied to business goals, revenue, and customer satisfaction. Yet, many project managers remain stuck in roles that reduce them to little more than "human calendars." They spend most of their time coordinating meetings, tracking tasks, and updating progress dashboards rather than shaping decisions that influence business outcomes. Why does this happen? Here are the root causes that keep PMs tethered to operations rather than strategy.

### 1. Tools That Automate But Also Limit

Modern project platforms—Jira, Asana, Monday.com—excel at automating task updates, reminders, and reporting. This often shifts the PM's role to one of real-time dashboard maintenance rather than strategic guidance. Academic research reveals that while tools improve efficiency, they often fail to fuel firm-level impact if PMs don't supplement them with meaningful interpretation[6].

A useful way to look at it: Tools handle what and when; PMs must focus on why these tasks matter and how they relate to business objectives.

## 2. Misaligned Priorities: Busy vs. Impactful

Many PMs fall into a pattern of equating activity with value: closing tasks, circulating minutes, and sending follow-ups. This "busy loop" keeps them visible—but not strategic. Executives care about impact: *Did this project increase customer retention, reduce costs, or support a new growth avenue?*

**A Common Case Scenario in Firms:** In many companies, project managers are often praised for their organizational efficiency—scheduling meetings flawlessly, ensuring timely updates, and keeping teams aligned. Yet, during critical discussions, such as budgeting sessions, these same PMs are frequently excluded. The reason? They rarely discuss ROI, cost-benefit trade-offs, or alternative paths that could influence strategic decisions. This is a common pattern observed across industries: operational excellence earns appreciation, but without demonstrating business insight, a PM's influence in decision-making remains limited.

---

[6] Maylor, "Beyond the Gantt Chart," *EMJ*

## 3. Rewarding Reports Over Recommendations

Organizational incentives play a powerful role. When leaders praise detailed status reports and precise timelines, PMs double down on those behaviours. But when visibility becomes the reward, strategic thinking becomes invisible.

Ask yourself: *When was the last time your update sparked a decision or a pivot, not just an acknowledgment?* If it was rarely, you might be reinforcing the "weeds trap" where depth of detail trumps strategic foresight.

## 4. Gaps in Business Fluency

Strategic conversations often revolve around budgets, opportunity costs, and ROI—a territory where many PMs feel unprepared. Lacking confidence, they retreat to scheduling and ticket tracking, ceding strategic ground to others.

This perpetuates a cycle: if PMs aren't fluent in business language, leaders won't involve them in high-level discussions. And without those experiences, PMs remain under-equipped for future clarity. Breaking this cycle starts with learning financial basics and practicing how to articulate impact in business terms.

## 5. Cultural and Structural Constraints

In many organizations, PMs are only brought in after key decisions are made. Once things are underway, PMs are expected to deliver—not to question scope, align strategy, or advise on alternatives.

Structural constraints like rigid hierarchies reinforce this through token involvement: required but not consulted. For PMs to break free, organizations must invite them earlier during ideation and goal-setting stages so they can influence priorities, not just execute them.

*What Separates Calendar-Holders from Strategic PMs?*

| Calendar PM | Strategic PM |
|---|---|
| Maintains dashboards and status reports | Translates status into business insights |
| Juggles tasks and schedules | Aligns deliverables with ROI and impact |
| Closes tickets | Recommends trade-offs and pathways |
| Able to explain *what* is done | Able to explain *why* it matters |

## Curiosity to Action: Turn the Tide

- Reflect on your last few status updates: Did you just report or advise?
- Identify three ways your next report can connect a metric (e.g., speed, cost, customer feedback) to a business goal.
- Seek out one meeting where you can gradually suggest alternative paths backed by data or business thinking.

Being perceived as a "human calendar" is not inevitable—it is a status you can outgrow. It starts with shifting your mindset from doing to influencing, from updating to guiding.

# What Organizations Actually Expect Now

The expectations for project managers have shifted dramatically. Companies are no longer satisfied with well-organized timelines and neatly closed task lists. They want project managers who can influence outcomes, make sense of complexity, and drive business value. The gap between what was once considered "good project management" and what executives expect today is widening—and those who fail to bridge it risk being left behind.

## 1. Business Alignment Above Everything

Executives now measure a project's success not by how closely it sticks to its original plan but by whether it advances organizational goals[7]. PMs who cannot clearly explain why a project matters to revenue, customer satisfaction, or competitive positioning struggle to gain influence.

A telling shift can be seen in how updates are received at leadership meetings. Progress charts and task completion percentages are quickly skimmed, but a statement like, *"This initiative is projected to reduce customer churn by 8% within the next quarter,"* holds attention.

To meet these expectations, PMs must start connecting their daily work to metrics leadership values—growth, profitability, market share, and customer outcomes. Asking early in the project lifecycle, *"What business problem are we solving, and how will we know we've solved it?"* positions PMs as partners rather than coordinators.

## 2. Outcome-Oriented Thinking, Not Timeline Obsession

A completed project that doesn't produce value is no longer seen as a success. Many organizations have learned this the hard way, delivering on time only to discover that the product or service does little to improve the bottom line.

Modern PMs are expected to think in terms of outcomes. That means assessing whether a feature, deliverable, or process

---

[7] "Beyond the Gantt Chart: Leadership Skills for Project Managers." *Bryan Mgt Services*, published May 2025. Accessed July 16, 2025. https://bryanmgtservices.com/project-managers/

genuinely drives the intended business result. It may even require recommending adjustments mid-project if the original plan no longer aligns with strategic priorities.

Consider this contrast:

- A traditional PM might report, *"We've completed 75% of the planned deliverables."*
- A strategic PM reframes it as, *"We've prioritized features that deliver 60% of the expected customer retention impact, ahead of schedule, allowing us to accelerate revenue recovery in Q3."*

The second statement speaks directly to leadership concerns, demonstrating outcome-based thinking.

## 3. Becoming Trusted Advisors

Executives are increasingly looking for PMs who can simplify complexity and guide trade off decisions. Being a trusted advisor means:

- **Challenging assumptions** when projects risk drifting from strategic goals.
- **Facilitating trade offs** by presenting clear scenarios, such as which option delivers the highest ROI or mitigates the greatest risk.
- **Communicating in business terms**, not technical jargon.

Leaders are more likely to involve PMs in strategic discussions when they consistently translate operational details into actionable insights. PMs who bring clarity rather than more data become indispensable, especially when decisions need to be made under uncertainty.

## 4. Case Scenarios: Proof from the Field

### *Case Scenario 1: Turning Progress Reports into Business Impact*

At a large manufacturing company, a project manager reviewed how much time was being spent on routine status meetings. By streamlining reporting and eliminating redundant updates, the team reduced meeting hours significantly, resulting in substantial cost savings for the company. Presenting these results in terms of dollars saved—rather than tasks completed—earned the PM credibility with senior leaders and secured them a role in future budget discussions.

### *Case Scenario 2: Influencing Priorities Through Business Insight*

In a software development environment, a project manager noticed that certain planned features would consume significant resources without adding meaningful customer value. Instead of simply flagging this as a risk, the PM prepared a comparative analysis, estimating which features could generate greater revenue and user engagement. By presenting this insight to leadership, the team shifted resources toward higher-impact initiatives. The PM was later invited to participate in strategic product planning sessions because of their ability to connect project decisions to business outcomes.

Thus, PMs who consistently demonstrate how their work influences revenue, cost savings, or customer experience are viewed as strategic partners, not just process facilitators. By translating project progress into meaningful business terms, they gain trust, visibility, and involvement in high-level decision-making.

# Chapter 2 – The High Cost of Staying in the Weeds

For years, "on time and on budget" has been the holy grail of project delivery. Project managers have been conditioned to measure success by adherence to timelines, scope, and cost. But that definition is increasingly outdated, in fact, dangerous in the modern business environment. Projects can meet every traditional metric, but they still fail the organization. They may be flawlessly executed yet disconnected from strategic goals, customer needs, or long-term impact. This gap between execution and relevance is where many project teams falter.

The problem is misplaced focus, not poor performance. Organizations risk delivering polished work that doesn't move the needle when the spotlight is fixed on processes instead of purpose. Executives are no longer impressed by progress charts or closed tickets. They want outcomes. And they're often frustrated when delivery excellence masks a lack of business impact.

This chapter will examine what happens when project teams stay buried in tactical work, losing sight of the bigger picture. You'll see how the weeds of daily execution, if not intentionally escaped, can trap not only PMs but entire organizations in cycles of misaligned output. We'll begin with the most subtle but costly issue of all: delivering perfectly and still failing.

## When Delivery Meets Disconnection

A dangerous myth continues to shape project culture: if you deliver everything on time and within budget, you've succeeded. It's no longer true. Organizations are filled with stories of projects that checked every box—completed tasks, hit milestones, stayed within budget—only to realize, months later, that nothing actually improved. No revenue was generated. No customer behaviour changed. No strategic goal moved forward.

In fact, research shows that globally, only 48% of projects are considered successful overall—including delivery on time, on budget, and delivering value—while 40% fall in a grey area, and 12% are outright failures[8]. Even more revealing: just 36% of organizations consistently deliver the full benefits of their projects, such as measurable customer or financial outcomes[9].

These data indicate that traditional delivery success metrics, such as schedule, scope, and budget, do not guarantee business impact. Project delivery can become hollow if PMs do not tie outcomes to strategy. Aligning project completion with measurable business benefits is essential for meaningful success.

## The Illusion of Progress

Overemphasis on execution visibility is one of the main reasons for the disconnect. Teams become hyper-focused on progress dashboards, burndown charts, and ticket velocity. These metrics create the illusion of momentum, but often fail to solve the right problem.

Consider a product launch that hits every delivery milestone. The features are complete, quality assurance signs off, and marketing is in sync. But two months post-launch, adoption is flat.

[8] "Project Management Institute Furthers Commitment to Maximizing Project Success." 2023. Pmi.org. 2023. https://www.pmi.org/about/press-media/2024/pmi-champions-a-new-era-of-project-success.

[9] "Project Management Statistics (2022) - Plaky." 2022. Project Management Hub | Plaky. April 1, 2022. https://plaky.com/learn/project-management/project-management-statistics/.

Customers didn't need those features. They needed a simpler onboarding experience—which was never scoped. Execution was flawless, but strategic alignment was absent.

This isn't an isolated scenario; it's becoming increasingly common, especially in large organizations with siloed functions and rigid frameworks. Delivery becomes mechanical. Reflection and course correction are sidelined. The work gets done, but not necessarily the right work.

## Why "More" Isn't Always Better

Another contributing factor is over-delivery—completing more than was initially planned in an attempt to add value. Ironically, this often results in the opposite. When teams add scope without re-evaluating the project's strategic objectives, they may increase complexity, stretch resources, and delay delivery of the very features that would have driven impact. In a value-driven framework, more does not always mean better—it can mean waste.

Executives are beginning to notice. In boardrooms, there is growing fatigue around project reports filled with operational detail but void of strategic insight. They're asking new questions: How does this help us enter a new market? Will it improve customer retention? What is the expected return? Many project managers are unprepared for these conversations—not because they lack competence, but because their performance has always been measured by different criteria.

## The Disconnect in Measurement

There's a fundamental misalignment between how teams measure success and what leadership actually values.

Let's look at it side by side:

| Execution-Focused Metrics | Outcome-Focused Metrics |
|---|---|
| % of tasks completed | Increase in customer satisfaction |
| Hours logged | Reduction in operational cost |
| Scope adherence | Time-to-market advantage |
| Budget variance | Growth in user acquisition |
| Team velocity | Impact on revenue or retention |

When teams optimize for the left column, they often lose visibility into the right. This is a strategic issue. If the PM and the project team aren't aligned with the business outcomes from the beginning, it's difficult to retroactively justify the work.

## The Cost of Irrelevance

Misaligned delivery has real consequences. It wastes resources, demoralizes teams, and erodes executive trust. Leaders begin to see project management as disconnected from strategy, reinforcing the outdated view that PMs are simply administrators. This is how the credibility gap grows—and how high-performing teams find themselves excluded from meaningful planning sessions.

Even worse, this misalignment ripples outward. When a project consumes time, attention, and budget without delivering clear value, it displaces other initiatives that might have. That opportunity cost is rarely calculated but is always felt in the long run.

## The Frustration Spiral

The danger of being stuck in the operational weeds isn't just inefficiency. It's frustration that corrodes decision-making, weakens leadership, and depletes teams. When project managers and leaders become overly focused on day-to-day details, they limit strategic visibility and unknowingly contribute to a culture where momentum is replaced by motion. Tasks are completed. Reports are updated. But progress stalls.

The frustration spiral is one of the most invisible, yet costly outcomes of poor role focus in modern organizations.

## When Leaders Drown in Detail, Progress Slows

Leaders who involve themselves in every decision, approve every slide, or question every execution step believe they're staying "close to the work." In reality, they are often clogging the flow of execution. As highlighted in Dame Leadership's insights, leaders caught in operational noise tend to become decision bottlenecks. The organization slows down not because people aren't working, but because they're constantly waiting for clarity, signoff, and permission.

What's worse is that this behaviour isn't isolated. It replicates. When executives micromanage, their teams begin to mimic that model, often assuming that busyness equals productivity. The result is a chain of management at every level addicted to staying visible—while strategic direction gets lost in the noise.

## The Emotional Toll on Project Managers

For project managers, this operational overload isn't just a time drain—it's an emotional burden. Juggling conflicting priorities, preparing never-ending reports, and managing daily task tracking leaves little room for strategic thought. Over time, this leads to disengagement and burnout.

According to Gallup's workplace data, employees who don't feel their work contributes meaningfully to broader goals are twice as likely to experience stress and dissatisfaction. The constant push into administrative detail can feel like a slow erosion of purpose for PMs who originally entered the field to drive innovation or cross-functional impact.

Over months—or years—this dynamic can silently push top performers out of organizations. Talented project professionals, capable of translating strategy into execution, leave not because they can't handle complexity, but because they're asked to manage busyness instead of outcomes.

## The Cost of Leading Without Elevation

Brilliance Within Coaching identifies a clear consequence of leaders remaining in tactical detail: energy depletion and poor delegation. Instead of enabling their teams to make autonomous decisions, they absorb responsibility for everything. This not only increases their own cognitive load but also creates role ambiguity for those under them.

No one is truly accountable for anything when everyone is responsible for everything.

Productivity drops, confusion spreads, and managers become firefighting machines, spending their time reacting rather than guiding. Meanwhile, strategic opportunities pass by unnoticed because no one has the bandwidth or perspective to recognize them.

This kind of leadership also leads to "false urgency"—where everything feels critical and nothing is prioritized. PMs are pulled in five directions at once, and teams lose their sense of focus. The organization may look busy from the outside, but energy is misdirected and drained inside.

## The Illusion of Productivity

One of the most damaging effects of this spiral is the illusion that things are going well. Progress reports are filled with updates. Boards are full of movement. Emails are flying. But the actual impact is minimal.

Researchers and practitioners often describe this as visible busywork, efforts made to demonstrate movement rather than create value. It is performative productivity, and while it may protect reputations or offer short-term validation, it often prevents meaningful outcomes from emerging.

In many teams, a PM's ability to produce a status report faster than anyone else becomes their "superpower," even though those reports rarely influence decisions. That level of efficiency masks a critical issue: when value is no longer at the centre of execution, tasks become vanity metrics.

## Signs You're Caught in the Spiral

PMs and team leads can often sense when they've been pulled into this cycle, even if they can't name it. Here are some warning signs that the frustration spiral has taken hold:

| Symptoms | What It Likely Means |
| --- | --- |
| Frequent rework after executive reviews | Lack of strategic alignment from the start |
| Status updates valued more than recommendations | Visibility prioritized over insight |
| Constant resource juggling without progress | Tactical focus overshadowing prioritization |
| Burnout across project teams | Lack of clarity, autonomy, or direction |

### Escaping the Cycle Starts with Awareness

The frustration spiral thrives in environments where execution is mistaken for effectiveness. Breaking it begins when leaders and PMs step back and ask: Are we doing the most valuable work, or just staying busy to stay visible?

It requires courage to say no to unnecessary detail, to question the way success is measured, and to shift conversations from tasks to impact. But without this shift, organizations will continue to deliver projects that check every box—except the one that matters most: business relevance.

## The Expectations vs. Reality Gap

A growing disconnect in many organizations is also largely due to executives and senior leaders asking for one thing, but project managers delivering something else. Projects still get delivered, but success feels incomplete, disconnected, and hollow.

At the core of this gap is misalignment between what leadership needs and what PMs believe is expected of them. This misalignment is the result of outdated role definitions, unclear expectations, and a culture that still equates visible activity with value creation.

### What Executives Expect

When surveyed or interviewed about what they truly need from their project managers, executives tend to highlight a consistent set of qualities:

- **Strategic alignment**: Can this PM connect the project's purpose to a broader business objective?
- **Trade off management**: When priorities collide or resources are strained, can this PM offer clear, defensible options?

- **Outcome focus**: Are we working toward impact, or are we just following a plan?
- **Executive clarity:** Can this PM communicate in a way that helps leadership make decisions faster?

In short, executives are looking for someone who can sit at the intersection of execution and strategic insight, a partner who brings clarity, not just coordination.

## What PMs Often Deliver Instead

The day-to-day reality inside many teams looks very different. Instead of strategic insight, what executives often receive includes:

- **Long weekly reports** filled with task-level updates
- Deep dives into **scheduling nuances and resource** availability
- **Reiterations of known risks** and pre-approved mitigation plans
- **Meeting recaps** that document everything, but drive nothing

This information is misaligned with what senior leadership values. Executives don't want to know whether the color-coded swim lanes are moving. They want to know whether the work is positioning the company for growth, stability, or change.

## Why This Gap Exists

Several factors are at play:

### 1. Outdated Role Definitions

While leadership expectations have evolved, many organizations have not redefined the PM role accordingly. Job descriptions still focus heavily on coordination, risk logging, and schedule management. There's little mention of financial acumen, stakeholder influence, or strategic framing.

As a result, PMs optimize for what their role technically demands, not for what the organization now needs. This leaves PMs stuck managing work, but not value.

## 2. Cultural Reinforcement from the Top

Leaders sometimes reinforce the very behaviours they're frustrated by. When executives focus on deliverable updates or demand hyper-detailed reports, they signal that executional detail is what matters most. Over time, this creates a feedback loop: PMs learn that surface-level visibility gets praised, and they stop pushing for deeper involvement.

What many executives fail to realize is that by staying entangled in details, they unintentionally discourage strategic thinking. If the senior team never makes space for strategic input from PMs, those PMs will continue to limit themselves to compliance, not contribution.

## 3. Lack of Confidence or Business Fluency

Even when PMs want to step up, many hesitate. Some fear they don't have the financial fluency or business vocabulary to communicate in leadership settings. Others believe it's "not their place" to question priorities or suggest pivots. Without support or coaching, these professionals default to the safety of reporting rather than risk bringing insight.

At its core, this expectation-reality divide erodes trust. PMs feel they're working hard but not being taken seriously. Executives feel they're not getting the strategic insight they need, even from seasoned project leaders. Over time, PMs are sidelined from planning conversations, further reducing their influence and deepening the divide.

More dangerously, this misalignment slows the entire organization. Decisions take longer. Priorities get blurred. Work continues, but momentum fades.

### Bridging the Gap

Fixing this isn't about retraining all PMs overnight—it's about clarity, support, and deliberate redefinition. Organizations must realign the PM role with strategic expectations. PMs must be invited into higher-level conversations and equipped with the tools to thrive there.

And most importantly, everyone, from executives to project teams, must stop mistaking activity for alignment. Only then can the PM evolve from task manager to true strategic partner.

## The Organizational Costs of Staying in the Weeds

When project managers and leaders spend their time buried in operational details, the cost extends far beyond a few inefficient meetings or delayed decisions. It ripples across the entire organization—affecting profitability, productivity, morale, and long-term agility. This is not an abstract idea. It's measurable, recurring, and preventable.

Let's stop pretending that tactical excellence always leads to strategic success. In reality, the longer a business remains focused on operational noise, the more vulnerable it becomes to missed opportunities, leadership fatigue, and organizational stagnation.

### The Financial Toll of Operational Obsession

Organizations that can't differentiate motion from momentum tend to waste time, money, and human energy. Studies estimate that poor leadership practices—including micromanagement, indecisiveness, and reactive planning—can cost businesses up to 7% of their annual revenue[10]. That's not just a rounding error. For mid-sized companies, this translates into millions lost to unproductive cycles and missed initiatives.

Furthermore, many companies experience a 5–10% productivity drag due to these behaviours, which better leadership could eliminate. Other impacts highlighted in the analysis include an opportunity for 3–4% gains in customer satisfaction and 1.5% revenue growth through improved leadership capacity[11].

## Productivity Without Purpose

Execution-heavy cultures often confuse busyness with progress. Everyone is working, everyone is reporting—but few are challenging priorities, revisiting assumptions, or driving change. The work gets done, but the reason behind it is forgotten.

This leads to what some researchers call strategic erosion: the gradual weakening of alignment between a company's current activities and its future goals. Over time, that misalignment compounds. Projects no longer reflect current business needs. Teams lose clarity. Leaders lose trust. The organization becomes slower in responding when it matters most.

One of the most dangerous outcomes of this erosion is decision paralysis. When everyone is focused on maintaining what's already in motion, no one stops to ask whether it should

---

[10] "The Cost of Poor Leadership on Your Revenue and Culture - Blog -." 2017. Gbscorporate.com. 2017. https://www.gbscorporate.com/blog/the-cost-of-poor-leadership-on-your-revenue-and-culture?

[11] Alises, Adriana. 2015. "The High Cost of Poor Leadership - John Spence." John Spence. October 6, 2015. https://johnspence.com/high-cost-poor-leadership/.

continue. Decisions get delayed. Risks get ignored. Pivoting becomes painful because no one is looking ahead—they're too busy looking down.

## Attrition, Fatigue, and Disengagement

The human cost of staying in the weeds is just as damaging as the financial one. PMs and team leads stuck in tactical cycles often report high levels of burnout. When the work is reduced to coordination—without context, impact, or trust—it feels mechanical. Engagement drops, creativity disappears, and turnover climbs.

This burnout doesn't only affect the individual. It affects the team. When capable PMs burn out or leave, their institutional knowledge walks out the door with them. New hires take time to ramp up—and often inherit the same structurally flawed roles that drove their predecessors away.

## Cultural Signals That Reinforce the Trap

Culture plays a critical role in either reinforcing or challenging this operational fixation. In many companies, the behaviours that get rewarded are not the ones that move the business forward. Detailed documentation, frequent check-ins, and reactive problem-solving are praised, while strategic questioning or simplifying complexity is seen as disruptive.

When a culture praises accuracy over insight and volume over value, it traps people in maintenance mode. Teams stop thinking like problem-solvers and start thinking like caretakers of status quo. This is how smart people, in strong organizations, slowly start producing work that no longer matters.

## Misalignment at the Top Goes Down the Line

When executives and PMs operate from different understandings of success, every layer beneath them is impacted. This project-executive alignment gap means that even if each team functions efficiently, the whole organization feels disconnected. The right hand doesn't know what the left hand is building—and both assume the other is driving strategy.

The longer this gap remains unaddressed, the more fragile the organization becomes. Teams develop habits based on internal optics rather than external impact. Leaders grow distant from customer value. And projects become polished artifacts rather than tools of transformation.

Thus, the cost of staying in the weeds is not just about lost time, but also lost influence. PMs who remain buried in execution become invisible at the decision table. Leaders who stay reactive lose credibility. Organizations that confuse progress with productivity become slower, weaker, and less adaptable over time.

Delivering efficiently and strategically are two different things. It's not enough to finish projects. Those projects need to solve real problems, support company priorities, and deliver visible results. If that connection isn't being made, the project falls short, no matter how well managed.

# Chapter 3 – The Tools Trap: When Processes Replace Purpose

Project managers today have access to more tools than ever before. From Jira and Asana to ClickUp and Trello, the project management landscape is rich with dashboards, notifications, timelines, and status trackers. These platforms promise clarity, alignment, and control. And in many ways, they deliver. Teams know what they're working on. Stakeholders receive regular updates. Tasks are visible. Progress is measured. But with all these gains in operational transparency, something essential often slips away: strategic purpose.

When project management becomes synonymous with "tool mastery," there is a risk of reducing the PM's role to that of a systems administrator rather than a business partner. Many project managers become highly skilled at keeping the tools up to date, automating workflows, and formatting reports. Yet, they struggle to connect project activity with business relevance. The illusion of control created by these tools can be seductive. It allows teams to feel productive without interrogating whether their efforts are moving the organization in the right direction. In this chapter, we dig into this overreliance on tools—why it happens, what it costs, and how PMs can break free from the trap of process over purpose.

## The Illusion of Control Through Tools

Project management software exists to solve a problem: the chaos of coordination. When used appropriately, tools like Jira, Asana, and ClickUp provide visibility, accountability, and a shared reference for distributed teams. But there's a tipping point. When tools begin to define the work rather than support it, teams can become trapped in a cycle of managing the system instead of driving meaningful progress.

## 1. Tool Visibility ≠ Strategic Clarity

Dashboards may reflect a well-organized sprint or show that 85% of tasks are complete, but they often fail to answer the question executives care most about: *"So what?" What problem did we solve? What outcome did we influence? Where is the ROI?*

A project management expert argues that executives don't want to see burndown charts. They want to understand how a project will reduce costs, increase revenue, improve customer satisfaction, or position the company competitively. Yet, when PMs rely too heavily on tools to communicate progress, they risk stripping away the narrative layer that provides strategic meaning.

*"The dashboard looks healthy, but we still don't know how this affects our Q4 objectives."*

This disconnect breeds frustration. Tools track activity. They do not, by default, interpret relevance.

## 2. Cognitive Offloading and the Loss of Intent

Cognitive offloading is a psychological phenomenon where people shift mental tasks onto external systems. In project management, this is what tools are designed to do. They help store, organize, and present data so the brain can focus on higher-level decision-making. However, the risk arises when teams start using tools as a substitute for critical thinking.

When every step is templated, every update automated, and every decision embedded into a pre-built workflow, PMs may stop asking essential questions:

- Why are we doing this project now?

- What strategic gap does it address?
- Is the current plan still aligned with business goals?

Tools reduce mental friction, but without reflection, they also reduce situational awareness.

## 3. Performance Theatre: The Danger of Looking Busy

A Jira board filled with green checkmarks looks impressive. So does a Gantt chart with smooth timelines and color-coded progress bars. But what happens when these visuals reflect execution without impact?

This is what researcher Cal Newport calls *"performance theatre"*: the presentation of busyness as a proxy for effectiveness. Teams may work hard and report frequently, but if those efforts are misaligned with strategic needs, they deliver little value.

Let's look at two scenarios:

| Scenario | Description | Outcome |
|----------|-------------|---------|
| PM A | Maintains a pristine Kanban board, delivers updates on time, keeps all tasks flowing | Stakeholders express gratitude but question the purpose of the work |
| PM B | Focuses on impact metrics, adjusts scope based on business needs, connects outcomes to revenue or risk | Stakeholders request input on strategic planning |

PM A might be celebrated internally by the organization. PM B becomes a trusted advisor.

## 4. Tool-Centric Thinking Narrows Influence

Tool expertise is often rewarded early in a PM's career. But over time, organizations seek leaders who can transcend systems and speak about opportunity, risk, and competitive advantage.

By focusing too much on mechanics, PMs risk narrowing their career trajectory. This is especially true when leadership conversations shift from "What did we build?" to "What did we enable?"

Executives operate in terms of:

- Strategic bets
- Portfolio trade offs
- Market timing
- Competitive differentiation

A PM who speaks the language of epics, sprint points, and ticket statuses will not influence those conversations, no matter how organized their boards are.

## 5. Tools Without Purpose = Noise

The issue isn't the tools. It's the default mode of using them as the end rather than the means. When teams equate motion with momentum, tools become distractions rather than enablers.

In summary:

| Tool Benefit | Risk When Overused |
|---|---|
| Visibility | False sense of progress |
| Structure | Inflexibility to pivot |
| Automation | Loss of critical questioning |
| Documentation | Noise without narrative |

Strategic PMs understand that tools are scaffolding, not the structure itself. To lead effectively, they must elevate the conversation—from metrics to meaning, from tasks to transformation. In the next section, we'll explore how relying solely on status updates further erodes credibility and why storytelling is the missing piece in strategic influence.

## From Status Reports to Strategic Storytelling

Walk into any project status meeting and you'll likely find a familiar tempo: a progress bar, a list of completed tasks, pending issues, and some charts pulled from a dashboard. For all the activities reported, stakeholders often leave the room and feel that they are no closer to understanding what the project actually means to the business. This disconnect is because of the lack of narrative.

Project management tools generate detailed visibility. Gantt charts, burn-down reports, and kanban boards illustrate throughput. But while this satisfies operational transparency, it doesn't answer the questions executives care about: *What's the business impact? Why should this project still matter? Are we still heading in the right direction? Status reports track the "what." Storytelling explains the "why."*

### The Shortcomings of Traditional Status Reporting

While tools are great for coordination, they strip away human context. Updates often reduce work to binary metrics: complete or incomplete. Milestones become just lines on a chart. The nuance of effort, trade-offs, or shifting priorities is lost.

Here's how conventional status updates fall short:

| Element | Typical Status Update | Strategic Storytelling |
| --- | --- | --- |
| Focus | Tasks and timelines | Value and outcomes |
| Language | Technical and static | Business-oriented and adaptive |
| Engagement | Low (informational) | High (emotional and rational) |
| Perspective | Present-focused | Future-anchored |

An update like "*Design phase 80% complete*" doesn't mean much to a CFO or CEO. But "*We've validated 3 key user flows that reduce onboarding friction by 40%, directly supporting our Q4 customer retention targets*" does. That's a narrative that links work to outcomes.

## Why Storytelling Resonates

Humans don't respond to charts alone. They respond to change framed as a journey. This is why strategic storytelling is now considered a core leadership skill. When PMs craft a clear narrative—where we are, where we're going, and why it matters—they invite decision-makers into a shared vision.

This narrative arc requires framing:

- **Current State:** What challenges or opportunities are we addressing?
- **Progress with Meaning**: What's been achieved—and how does it tie to goals?
- **Upcoming Moves:** What's next, and what value does it unlock?

Effective PMs now use storytelling as a bridge between execution and strategy. This helps stakeholders understand the status of work and its strategic relevance. In executive settings, the ability to convey this narrative has become a key differentiator.

## Case Scenario: Narrative over Numbers

At a healthcare technology company, the PM leading an infrastructure overhaul stopped using tool-driven dashboards in executive meetings. Instead, she framed updates using narratives: *"With our new system migration 60% complete, we've already seen 18% faster claims processing, allowing patient queries to resolve a day earlier on average. This is directly impacting patient satisfaction scores."*

The outcome? Leadership shifted from being passive recipients of updates to active supporters, greenlighting additional investments. The narrative framed execution as strategic momentum.

## Strategic Narratives as Decision Tools

Beyond engagement, stories help simplify complexity. In large-scale projects, role confusion, shifting priorities, and misalignment are common. Without a unifying story, every department interprets progress through its own lens. Strategic narratives act like compasses, anchoring teams to a shared purpose.

When everyone understands the why, alignment improves:

- Teams self-correct based on intent, not just instructions.
- Trade-offs are evaluated through a strategic lens.
- Leaders avoid micromanaging because they trust the direction.

*What Makes a Strategic Narrative Work?*

- **Clarity over Jargon**: Use business language, not just project lingo.

- **Relevance to the Audience:** Frame updates in terms stakeholders care about—growth, cost, risk, customer experience.
- **Continuity**: Create a thread between past, present, and future.
- **Actionable Insight**: Don't just tell a story—guide a decision.

### Building the Habit

Most PMs already hold the raw ingredients for storytelling: project insights, progress data, user feedback, and business metrics. What's missing is the framing. Instead of ending your update with *"We're 50% done,"* consider ending with, *"Here's what that 50% unlocks for our business."*

Making this shift doesn't require more meetings. It requires changing the message. Replace the need to prove busyness with the desire to communicate value. Storytelling gives purpose to the data.

## Building Business Visibility Through Narrative

In complex organizations, visibility is currency—but not all visibility holds the same weight. While dashboards and ticketing systems broadcast activity, they don't always communicate value. Leadership doesn't just want to know that work is being done; they want to understand what that work means. And that's where narrative becomes a strategic tool.

When project managers learn to tell stories—not fairy tales but data-backed, business-anchored stories—they shift from task trackers to meaning-makers. Strategic storytelling doesn't just replace a slide deck; it reshapes how leaders perceive the PM's role.

## Data Alone Doesn't Move Minds

PMs often assume that sharing progress metrics—tasks completed, milestones met, tickets closed—is enough to demonstrate success. But this kind of status reporting rarely answers the leadership's real question: So what?

A Harvard Business Review article explains that executives need context to act. In large-scale projects, decisions aren't made based on numbers alone. They're made when those numbers are tied to business outcomes: customer impact, competitive advantage, risk mitigation, or growth opportunities. A PM who can bridge that gap—who can say, "We're not just 80% done; we've removed the friction that was delaying $3M in revenue this quarter"—earns a seat at the table.

## Strategic Narratives Drive Alignment

Strategic narratives are not fluff. They are intentional stories crafted to:

- Articulate the why behind the work
- Map current progress to future outcomes
- Connect the dots across departments, timelines, and goals

These narratives help stakeholders see the long game. For instance, when launching a new internal CRM, a PM saying, *"This system will improve support response times by 40%, reducing churn and lifting NPS,"* is far more persuasive than *"We're 3 sprints away from full integration."*

Narratives like these shift the conversation from status updates to value creation and clarify roles. As the Institute of Project Management notes, when stories are well-told, ambiguity fades.

People know what matters, why it matters, and how their part contributes.

## Visuals Amplify Memory and Engagement

Visual storytelling—using charts, customer journeys, or even simple diagrams—helps people retain and respond to information. According to experts, executives exposed to visual narratives were 30% more likely to recall project implications and make quicker decisions in steering meetings.

Imagine two updates on a delayed feature:

- A PM shows a slide with the status bar stuck at 65%.
- Another PM shows a customer journey, highlighting how the missing feature leads to a 3-day delay in user onboarding—and overlays projected churn risk if it's not resolved within two weeks.

Which version is more likely to spur executive action?

Storytelling isn't just communication. It's influence.

## The Case for Value Visibility

In one case study from Celoxis, a consulting firm revamped its reporting strategy. Instead of weekly slide decks filled with status bullets, PMs were trained to present progress using client impact narratives. They started each review meeting with a question: "What's different in the client's world this week because of our work?"

The result? Executive engagement spiked. Issues were resolved faster because they were framed through outcomes, not tasks. Within one quarter, project delivery speed improved by 17%, and internal satisfaction surveys showed a 28% jump in stakeholder trust.

The takeaway: When PMs communicate in terms of value delivered—not just effort spent—they are no longer seen as administrators. They're seen as strategic partners.

*Summary Table – Tools vs. Storytelling*

| Tool-Centric Approach | Storytelling-Centric Approach |
|---|---|
| Dashboards, tickets, progress bars | Narratives of impact, decisions, outcomes |
| Monitoring work, not meaning | Connecting work to business purpose |
| Passive updates | Engaged, emotionally resonant conversations |
| Operational compliance | Strategic alignment and clarity |
| Activity-focused | Outcome-oriented |
| Task granularity | Business relevance and cohesion |

In this chapter, we've seen how storytelling transforms the role of the PM: from executor to interpreter. From managing motion to translating momentum. Tools are necessary, but they're not sufficient. Visibility alone doesn't build trust; value clarity does.

As we turn the page, we'll explore what it takes to move from reactive orchestration to proactive leadership. Because in the evolving world of work, being seen is no longer enough. It's time to be heard, strategically.

# Part II

# The Shift: Becoming a Strategic Partner

# Chapter 4 – Thinking Like a Business Leader, Not a Task Owner

Many project managers excel at execution—tight timelines, controlled budgets, and efficient task management. Yet these strengths rarely impress senior leadership. Executives want to know why the work matters, how it impacts the market, and whether it supports strategic goals.

The real shift comes when PMs think like business strategists, not schedulers. Effective leaders view projects within a wider ecosystem of customer expectations, competition, regulations, and market shifts. They measure success by outcomes and value created.

Adopting this mindset means blending operational discipline with entrepreneurial vision—being curious about the business model, fluent in ROI and risk, and willing to challenge weak strategic cases. This chapter explores how PMs can move beyond task ownership to lead with a business leader's perspective.

## The Mindset Shift – From Execution to Enterprise Thinking

### Why This Shift Matters

In traditional project management, success is often defined by the "iron triangle": scope, time, and cost. Deliver within these boundaries, and you're deemed effective. However, meeting these parameters is no longer enough in a business environment defined by rapid change. A project can be delivered on time and on budget yet fail to create measurable business value. Thinking like a business leader means redefining success around strategic impact, whether the initiative moves the company closer to its goals, strengthens its market position, or delivers tangible benefits to customers.

## Contrasting the Two Mindsets

A useful way to visualize this shift is to compare the characteristics of a task owner and a business leader.

| Task Owner | Business Leader |
|---|---|
| Focuses on schedules, tasks, and outputs | Focuses on vision, market impact, and measurable value |
| Measures success by completion | Measures success by outcomes |
| Works within given requirements | Challenges and refines requirements for strategic fit |
| Sees stakeholders as information sources | Sees stakeholders as |

While both roles require discipline, the business leader mindset adds strategic curiosity—asking not just what needs to be done but why and to what effect.

## Lessons from Startup Founders and Product Owners

Harvard Business Review research shows startup founders thrive on agility, market awareness, and adaptability—constantly testing assumptions and pivoting to seize opportunities. Product owners adopt a similar mindset, prioritizing initiatives that deliver maximum customer and business value, even if it means discarding approved work.

Thinking like a CEO demands the same breadth: understanding the full value chain, from idea to revenue or brand equity. For project managers, this means looking beyond departmental boundaries to link their work with marketing, finance, customer experience, and post-launch outcomes.

## From Task Metrics to Business Metrics

Many PMs are measured by operational indicators—task completion rates, defect counts, milestone adherence. These are important, but they tell only part of the story. Business leaders look at metrics such as:

- **Revenue impact** – How will this project drive top-line growth?
- **Customer retention** – Will it improve loyalty or reduce churn?
- **Operational efficiency** – Will it lower costs or reduce time-to-market?
- **Risk reduction** – Does it protect the company from regulatory, reputational, or operational threats?

By incorporating these measures into project tracking, PMs signal to executives that they are thinking in business terms, not just delivery terms.

## Practical Mindset Reframes

Shifting from execution to enterprise thinking involves deliberate changes in the questions you ask, the language you use, and the priorities you set.

| Instead of asking... | Ask... |
| --- | --- |
| "What's the next task?" | "What's the next most valuable outcome?" |
| "Are we on time and on budget?" | "Are we on purpose and on impact?" |
| "Have we met requirements?" | "Do these requirements still serve the business objective?" |

These reframes change how you engage with stakeholders. They move conversations from tactical check-ins to strategic dialogues.

## Building Credibility Through Strategic Curiosity

Enterprise thinking isn't about abandoning detail management—it's about embedding it within a broader narrative of value. A PM who can explain how a change request will influence market positioning or how a delay might affect quarterly revenue speaks in terms that leadership understands. Over time, this positions the PM as a trusted adviser rather than a process administrator.

## A Real-World Illustration

Consider a software PM overseeing a compliance upgrade. The task owner would focus on ensuring the release meets legal requirements by the deadline. The business leader PM would also explore how the compliance upgrade could be leveraged for competitive advantage—perhaps by marketing the product as *"security-certified"* ahead of rivals, creating a customer trust differentiator.

This mindset turns a necessary operational project into a strategic win, aligning execution with brand growth.

*Key Takeaways from the Shift*

- Operational excellence is the foundation, but strategic influence is the differentiator.
- Business leader PMs anticipate market shifts and adjust priorities accordingly.
- Speaking the language of business builds influence with executives and cross-functional leaders.

By adopting this mindset, project managers expand their role from managing deliverables to shaping the future of the organization.

## Discovery → Direction → Delivery

The transition from task owner to business leader is a matter of process. Strategic project managers follow a value-driven sequence: Discovery → Direction → Delivery. This approach ensures that what gets executed is both strategically relevant and positioned to create measurable impact. It replaces the default *"start executing as soon as requirements arrive"* model with a deliberate process that prioritizes understanding before action.

### 1. Discovery: Understanding the Landscape Before Acting

Many projects are greenlit without adequate exploration of the problem space. The project manager is handed a set of requirements and told to "make it happen," yet these requirements often reflect assumptions rather than validated needs. A Harvard Business Review article argues that leaders who think like startup founders actively test market hypotheses, validate customer needs, and investigate the competitive landscape before committing resources[12].

In this stage, the PM acts less like a scheduler and more like a business analyst, strategist, and even anthropologist. Core activities include:

---

[12] Ashkenas, Ron. 2023. "Project Managers Should Think like Startup Founders." Harvard Business Review. November 2, 2023. https://hbr.org/2023/11/project-managers-should-think-like-startup-founders.

- **Stakeholder Mapping** – Identifying not just the decision-makers, but also influencers, end-users, and hidden stakeholders who can impact adoption.
- **Market and Competitive Scanning** – Using internal data, market reports, and industry trend analysis to validate whether the initiative solves a meaningful problem.
- **Business Case Alignment** – Ensuring the project aligns with measurable strategic goals such as revenue growth, cost reduction, customer retention, or regulatory compliance.

Experts note that this early investigative work mirrors a CEO's approach: understanding the full value chain and how the initiative will contribute to the organization's bottom line[13]. Without this clarity, PMs risk delivering polished outputs that solve the wrong problem.

## 2. Direction: Shaping the Strategic Path

Research also confirms the need and value of the initiative, the PM's role shifts to defining a clear direction—the "*why*" and "*how*" of execution. At this point, the project manager functions like a product owner, translating insights into a roadmap that stakeholders can rally behind.

Research emphasizes that the product owner mindset is about prioritizing outcomes over outputs[14]. This involves:

[13] Isiakpona, Dennis. 2025. "Project Managers vs. CEOs: More Similar than You Think Great CEOs Don't Just Manage—They Lead, Strategize, and Drive Results." Linkedin.com. February 26, 2025. https://www.linkedin.com/pulse/why-project-managers-must-think-like-ceos-dennis-isiakpona-vnywe/.

[14] Overeem, Barry. 2016. "Project Mindset or Product Mindset?" Medium. The Liberators. March 23, 2016. https://medium.com/the-liberators/project-mindset-or-product-mindset-8d54e87a3f6d.

- **Defining Success Criteria in Business Terms** – Moving beyond "launch by Q4" to "increase customer onboarding rate by 15% within six months."
- **Balancing Strategic and Tactical Goals** – Ensuring operational milestones are tied to broader business outcomes.
- **Challenging Misaligned Requirements** – If a feature or deliverable does not serve the core business objective, it should be questioned or removed.

The Project Mindset vs. Product Mindset framework reinforces that direction-setting requires active trade-off decisions. Instead of trying to deliver everything, the PM ensures the most valuable components are prioritized and resourced first.

This stage is also where role clarity becomes critical. A *ScienceDirect* study on project delivery management (2023) warns that unclear ownership and decision-making authority in large projects often leads to duplicated efforts and wasted resources[15]. The PM must establish governance early—who approves changes, who owns risks, and who communicates with stakeholders.

## 3. Delivery: Executing with Strategic Intent

Delivery is the stage most PMs are comfortable with, but in this model, execution is not just about "getting things done." It is about protecting strategic intent during the chaos of delivery.

---

[15] Rehan, Ashok, David Thorpe, and Amirhossein Heravi. 2024. "Project Manager's Leadership Behavioural Practices – a Systematic Literature Review." Asia Pacific Management Review 29 (2). https://www.sciencedirect.com/science/article/pii/S1029313223001057.

Research highlights that delivery should be continually informed by feedback loops—market signals, stakeholder sentiment, and evolving business priorities[16]. This requires the PM to maintain agility, pivoting if new data reveals a better path to value.

Practical strategies for strategic delivery include:

- **Frequent Value Checkpoints** – Regularly revisiting the business case to ensure continued relevance.
- **Story-Driven Reporting** – Using narrative dashboards that explain not only progress but also implications for business outcomes
- **Visual Decision Aids** – Incorporating charts, infographics, and scenario maps to help executives quickly grasp trade-offs.

A case study illustrates this well. A technology company replaced traditional milestone reporting with impact-driven narratives supported by visuals. By showing executives why a shift in scope would deliver greater market impact, the PM secured faster approvals and ultimately delivered ahead of schedule, earning stronger executive trust.

## Why the Sequence Matters

Following **Discovery → Direction → Delivery** creates resilience against the common trap of "*execution without impact*." Starting with delivery risks wasted investment if the foundation is flawed. Conversely, discovery and direction ensure that the work being delivered is strategically sound, relevant, and understood by stakeholders.

---

[16] Watson, Elise. 2024. "Leveraging the Power of Data Storytelling for Project Managers." Clarkston Consulting. September 13, 2024. https://clarkstonconsulting.com/insights/data-storytelling-for-project-managers/.

This sequence mirrors the thinking patterns of CEOs and product leaders:

- They discover market opportunities before committing capital.
- They set a direction that aligns the organization's capabilities to those opportunities.
- They deliver with relentless focus on value creation, not just completion.

## Key Behaviors for PMs Applying This Model

| Stage | Business-Leader Behaviors | Impact |
|---|---|---|
| Discovery | Ask strategic questions, validate assumptions, map stakeholders, analyze market forces | Avoids executing irrelevant or low-impact projects |
| Direction | Define business-driven success criteria, challenge misaligned requirements, prioritize high-value features | Ensures resources are focused on the highest-value work |
| Delivery | Maintain agility, protect strategic intent, communicate through narratives and visuals | Builds executive trust and accelerates decision-making |

By mastering this sequence, project managers elevate themselves from coordinators of work to navigators of business value. This deliberate flow—from curiosity to clarity to committed action—turns projects into strategic assets rather than operational expenses.

## Business Acumen in Practice – Decisions that Shift the Game

The difference between a competent project manager and a strategic business leader often comes down to how decisions are made when there's no obvious playbook. Tools, schedules, and workflows keep projects moving, but the real value comes from the judgment calls that protect the organization's resources, seize unexpected opportunities, and strengthen its competitive position. This is where business acumen shifts a PM's role from managing deliverables to shaping outcomes that matter to the enterprise.

## Core Business Skills Every Strategic PM Should Cultivate

While leadership style and industry knowledge are important, certain business skills form the backbone of high-impact decision-making.

### *1. Financial Literacy*

Understanding project finances goes beyond staying within budget. Strategic PMs can:

- **Calculate Return on Investment (ROI)** to ensure initiatives deliver measurable value.
- **Assess the Payback Period** to determine how quickly the investment recoups.
- **Factor in Total Cost of Ownership (TCO),** considering both upfront and ongoing costs.
- **Evaluate Opportunity Cost** to compare potential benefits of alternate uses of resources.

When a PM can articulate the business case in financial terms, executives take notice. This language connects project performance directly to the organization's balance sheet.

### 2. Market Awareness

Business-minded PMs monitor customer sentiment, competitor moves, and regulatory shifts. This isn't about becoming a market analyst, but about integrating external signals into decision-making. For instance, if new technology adoption in the industry accelerates, a PM might expedite related features to keep the company ahead of the curve.

### 3. Risk–Reward Thinking

Traditional project risk registers focus on mitigation—avoiding harm. Strategic PMs also weigh the potential upside. A calculated risk might open a new revenue stream, strengthen market share, or speed delivery. Balancing potential loss against measurable gain is a hallmark of mature decision-making (Harvard Business Review, 2023).

### 4. Change Leadership

Projects rarely unfold without disruption—whether due to market changes, internal restructuring, or stakeholder shifts. Strategic PMs lead through these moments with transparency, aligning teams to a renewed vision rather than allowing momentum to erode.

## Example Scenarios – Where PM Decisions Changed the Trajectory

A business-minded project manager stands out not just for delivering work, but for making the kind of decisions that redirect a project toward higher value. The following scenarios illustrate how such choices can shape both project and business outcomes:

### 1. Redirecting Resources Based on Market Signals

Midway through a software development project, the PM notices emerging competitor offerings that make a planned feature less appealing to customers. Rather than push forward simply because it was in scope, they engage stakeholders to reassess priorities. The team reallocates budget and time toward a more in-demand capability, protecting market relevance and preventing sunk-cost waste.

### 2. Turning Compliance into Competitive Advantage

In a supply chain project driven by new regulatory requirements, the PM spots an opportunity to integrate sustainability measures beyond compliance. By involving marketing and CSR teams early, the project shifts from being a cost obligation to becoming a brand-strengthening initiative, improving public perception and customer loyalty.

### 3. Protecting Value Through Scope Discipline

During an infrastructure upgrade, stakeholders propose adding a range of non-critical features. The PM conducts a cost–benefit discussion with the leadership team, demonstrating how the additions could dilute resources and delay core functionality. By keeping focus on the high-impact deliverables, they ensure the project meets essential business objectives without overextending the budget or timeline.

### 4. Anticipating Change and Leading Through It

In a global rollout, geopolitical events disrupt supply chains. Instead of waiting for formal risk reviews, the PM proactively brings procurement, logistics, and finance together to rework the delivery model. The shift minimizes disruption and keeps customer commitments intact, reinforcing leadership trust in the PM's strategic value.

### Why These Decisions Stand Out

These examples work because they demonstrate:

- **Active environmental scanning** – PMs aren't passive recipients of scope changes; they initiate conversations when conditions shift.
- **Business-first reasoning** – Decisions are framed in terms executives understand: profitability, competitive positioning, customer value.
- **Influence beyond formal authority** – Strategic PMs win support by aligning their recommendations with broader business goals.

When PMs operate with this level of business acumen, they stop being viewed as cost centers and start being recognized as growth enablers.

Business-minded PMs blend operational discipline with commercial insight. They ask:

- *Does this initiative still serve our strategic priorities?*
- *What external factors could derail—or accelerate—our success?*
- *If we invest here, what are we giving up elsewhere?*

These questions lead to decisions that protect resources, sharpen focus, and open new opportunities. Over time, this shifts the PM's reputation from someone who delivers projects to someone who drives the business forward.

The evolution from task ownership to business leadership is a complete reframing of the project manager's role. Strategic PMs don't just follow the delivery plan; they help shape the business plan. They understand the financial stakes, read market signals, and influence decisions that move the organization forward. By doing so, they escape the narrow confines of schedules and deliverables and step into the leadership arena where purpose, value, and impact define success.

This is the essence of thinking like a business leader: you're no longer a messenger of progress updates, but a trusted partner in steering direction. Your seat at the table is earned through the clarity and insight you bring to decisions that matter.

# Chapter 5 – Always Start With "Why Does This Matter?"

In high-performing organizations, projects are measured by the business value they create. This is why strategic project managers start every conversation by asking, "Why does this matter?" The answer connects day-to-day activities to outcomes executives care about: growth, profitability, competitive advantage, and customer satisfaction. Without this connection, even technically flawless projects can fade into obscurity, failing to influence the company's direction or bottom line.

Let's explore how to reframe your role from executor to business partner, linking initiatives directly to the organization's strategic objectives. You will see how adopting a strategic lens changes decision-making, improves stakeholder alignment, and elevates your credibility as a leader who drives results that matter.

## The Strategic Lens – Linking Projects to Business Objectives

### From Outputs to Outcomes

Too many projects are framed around outputs: features built, reports submitted, or systems implemented. But as research points out, strategic influence in project management comes from demonstrating how these outputs lead to outcomes that matter to the business[17]. An upgraded CRM, for example, is not

---

[17] RMC Learning Solutions. 2025. "Business Acumen for PMs: From Project Execution to Strategic Influence - RMC Learning Solutions." RMC Learning Solutions. May 24, 2025. https://rmcls.com/learn/blog/business-acumen-for-pms-from-project-execution-to-strategic-influence/?srsltid=AfmBOoopgwyY_x3Hx0uGiKMmCSopqylM4F67u8LqgO7Bl2_XdtNoC7xq.

just a software deployment, but a tool for improving lead conversion rates and customer retention.

Shifting from outputs to outcomes means:

- Translating deliverables into business metrics: Instead of reporting *"feature X delivered,"* communicate *"feature X reduced processing time by 40%, saving $500K annually."*
- Defining value early: Before execution, identify what measurable improvement the project should produce— cost reduction, increased sales, reduced churn, improved compliance.
- Aligning metrics to strategy: If the company's priority is entering a new market, demonstrate how the project supports market entry, not just operational improvements.

A strategic PM starts by mapping each deliverable to a specific organizational goal:

| Deliverable | Business Objective | Measurable Outcome |
|---|---|---|
| CRM upgrade | Increase sales conversion | +10% in closed deals within 12 months |
| Supply chain redesign | Reduce operational costs | $1.5M annual savings |
| New mobile app feature | Improve customer engagement | +15% monthly active users |

This requires a change in mindset and familiarity with the organization's mission, vision, and KPIs. Without that connection, even a technically flawless project can be seen as irrelevant noise. For example, a retail chain's inventory management upgrade could be framed not as an IT success, but as a way to cut stockouts by 25%, directly improving sales during peak seasons.

A project charter that lists objectives without tying them to the organization's mission or KPIs risks being seen as a side activity rather than a core growth driver. Strategic PMs:

- Study the corporate strategy before project kickoff.
- Map each deliverable to at least one strategic pillar— e.g., "Enhancing customer experience" or "Driving operational excellence."
- Use language familiar to executives. If leadership measures "customer lifetime value," frame project benefits in those terms.

Research emphasizes strategic and business management as one of the three core competency areas[18]. Applying this principle, PMs act as translators between technical execution and business vision, ensuring every stakeholder sees the direct relevance of the work.

## Understanding the Business Terrain

Before execution, strategic PMs investigate the landscape their project will operate in. This practice is supported by research showing that market awareness and business acumen are critical factors in project success[19]. This means answering:

---

[18] Brown, Lucy. 2025. "What Is the PMI Talent Triangle for Project Management? A Detailed Guide." Invensis Learning Blog. July 11, 2025. https://www.invensislearning.com/blog/pmi-talent-triangle-for-project-management/.

[19] "(PDF) the Impact of Project Management in Achieving Project Success-Empirical Study." n.d. ResearchGate.

- How will this initiative impact revenue or cost efficiency?
- Does it align with current market conditions and competitive threats?
- Will it enhance customer experience in a measurable way?

For instance, a product launch timeline shouldn't be set solely by development readiness—it must consider seasonal market demand, competitor release schedules, and emerging technology trends. Ignoring these external factors is like navigating without a map.

## Framing the "Why" Through Strategic Communication

The way a project is presented matters as much as its technical merit. The Pyramid Principle[20] is a powerful tool for framing project pitches:

- Start with the answer first – The high-level business value.
- Support with key reasons – Strategic benefits tied to company priorities.
- Back with evidence – Data, market analysis, or pilot results.

Example: Instead of pitching a "system migration" as an IT necessity, reframe it as a customer retention and revenue

https://www.researchgate.net/publication/330480266_The_impact_of_proje ct_management_in_achieving_project_success-_Empirical_study.

[20] MacKay, Jory. 2022. "The Pyramid Principle - How to Effectively Pitch Projects." Planio. May 10, 2022. https://plan.io/blog/pyramid-principle-pitching/.

protection initiative, explaining how outdated systems are causing service delays that could drive churn. This shift in narrative changes the perception from a cost center to a profit-protecting investment.

## Business Acumen as a Project Driver

Research emphasizes understanding profit drivers, brand positioning, and customer expectations before setting execution plans. But business acumen is more than financial literacy and involves the ability to see where the organization is headed and how a project can accelerate that journey.

Practical applications include:

- **Cost–Benefit Analysis**: Evaluating trade-offs to prioritize high-return activities.
- **Market-Driven Prioritization:** Selecting features or initiatives that respond to evolving customer needs.
- **Stakeholder Value Mapping**: Identifying what each decision-maker values most and tailoring communications accordingly.

## Case in Point – Reframing to Gain Buy-In

*Atlassian – measurable latency reduction tied to customer experience*

Atlassian moved storage for key applications to Amazon FSx for NetApp ONTAP[21]. Average application response times improved by 17%, which the company links to better user productivity and

---

[21] "Atlassian Reduces Latency by 17% and Saves $2.1 Million with Amazon FSx for NetApp ONTAP | Atlassian & NetApp Case Study | AWS." 2025. Amazon Web Services, Inc. 2025. https://aws.amazon.com/partners/success/atlassian-netapp/?.

customer experience. This is exactly the kind of "reframe" leadership responds to: not tech upgrade, but "faster experiences that retain and delight users.

# Communicating Value Through the Pyramid Principle

## Why the Pyramid Principle Matters for Strategic PMs

When a project manager presents to senior leaders, the challenge is basically the complexity of the message, not the complexity of the work. Executives deal with an overwhelming stream of information every day. Their time is limited, their attention fragmented, and their priorities tied to business-critical decisions. The Pyramid Principle addresses this reality by front-loading the answer or key point, then supporting it with layers of evidence.

In project management contexts, this means you don't start by detailing the timeline, budget, or Gantt chart. You start by answering the one question an executive unconsciously asks in every meeting: "Why does this matter to the business?"

The principle's power lies in reducing cognitive load. Studies in business communication show that when audiences hear the key conclusion first, they retain more of the detail that follows because their brains have a framework to organize it. For PMs seeking to influence at the strategic level, this is a decisive advantage.

## Starting With the "Why" in Pitches

Instead of walking stakeholders through a chronological report of what has happened, focus on the end-state benefit and link it directly to organizational priorities.

Example:

- Weak pitch: "We've completed 60% of the migration tasks, integrated the new APIs, and trained half the support team."
- Strong pitch: "We're on track to reduce average customer onboarding time by 25% this quarter, which directly supports our goal of increasing net retention by 10% this year."

That first line instantly answers "Why does this matter?" and connects the project to revenue growth, something every executive understands and cares about. Only after this point do you layer in the technical or operational details.

Research highlights that presenting your main point first increases decision-making speed because the audience doesn't have to piece together the conclusion from scattered facts. They're primed with the "why" before they hear the "how."

## Structuring the Message

The Pyramid Principle works best when you think of your communication in three layers:

1. *Outcome First (Apex of the Pyramid)*

This is the "headline" for your project update or proposal. It should be concise, measurable, and directly linked to a strategic driver such as growth, efficiency, or risk reduction.

## 2. Impact Level (Middle Layer)

This explains why the outcome matters in context—tying it to customer satisfaction, market positioning, compliance, or cost control. This is where you show alignment with KPIs and the organization's mission.

## 3. Detail Level (Base Layer)

This is where operational facts, milestones, and risks live. At this stage, the executive has already bought into the "why," so they are more receptive to absorbing technical details.

Here's a simplified illustration:

| Pyramid Layer | Purpose | Example for a Cybersecurity Upgrade |
|---|---|---|
| Outcome | State the business value | "This upgrade will cut potential downtime risk by 40%, protecting $12M in annual revenue from disruption." |
| Impact | Explain strategic alignment | "Our current security gaps could trigger compliance fines and erode client trust—two risks directly linked to customer churn." |
| Detail | Provide supporting facts | "We've deployed new threat detection software to two-thirds of endpoints and completed 85% of penetration testing." |

## Framing in Business, Not Technical, Language

Even when a project is inherently technical, executives think in terms of business impact:

- Growth (market share, customer acquisition, revenue)
- Efficiency (time saved, cost reductions, resource optimization)

- Risk (regulatory compliance, operational continuity, brand reputation)

For example, instead of describing a system upgrade in terms of "faster database queries," frame it as "reducing average transaction processing time by 40%, enabling customer service to handle 20% more cases per hour."

This shift in language ensures your audience sees your project as an investment in business outcomes, not just a cost center.

## The Practical Tool: One-Minute Elevator Pitch

A practical way to embed the Pyramid Principle into your daily project communications is to develop a **"Why → What → How → Next Steps"** elevator pitch.

Template:

- Why — Start with the strategic outcome or value proposition.
- What — Briefly describe the project or initiative in business terms.
- How — Outline the approach or key enablers that will make it successful.
- Next Steps — Clarify immediate actions or decisions needed.

*Example – Marketing Automation Project:*

- **Why**: "We aim to increase qualified leads by 30% over the next two quarters, which supports our annual revenue growth target."
- **What:** "This project integrates our CRM with an AI-driven marketing platform to segment and nurture leads more effectively."

- **How**: "We'll automate follow-up sequences, personalize messaging based on user behavior, and improve response times."
- **Next Steps**: "We need budget approval for the final integration phase by the 20th to hit our Q2 launch."

Research emphasizes that a well-crafted elevator pitch allows you to communicate value to any audience—even if you only have 60 seconds of their attention—by prioritizing the "why" and sequencing information logically.

## Advanced Application: Tailoring the Pyramid for Different Stakeholders

Not all executives care about the same outcomes. The CFO may focus on cost savings and ROI, while the CMO may care more about market penetration or brand perception. The Pyramid Principle allows you to adapt the "Outcome" layer for each stakeholder without changing the core facts.

*Example – Same Project, Different Stakeholders:*

- **CFO:** "The initiative is projected to reduce operating expenses by 8% within the first year."
- **CMO**: "This will cut campaign cycle time by 25%, allowing us to respond faster to market trends."

By reframing the top of the pyramid for each audience, you increase the likelihood of buy-in across the leadership spectrum.

*Why This Works in the Real World*

The Pyramid Principle aligns with how people process and retain information. Neurological studies on decision-making show that framing information around a central conclusion activates brain regions associated with goal-directed thinking, making the audience more receptive. In practical terms, it means your project updates won't just be heard—they'll be acted on.

A PM who applies this consistently will find that meetings shift from status recitations to strategic discussions, where stakeholders contribute ideas, remove roadblocks, and champion the project's success.

## Converting Stakeholder Interest into Sustained Momentum

Getting initial buy-in from stakeholders is often only the beginning. The real challenge lies in keeping that interest alive and using it to fund momentum through to project completion and beyond. Here, we explore how project managers can transform fleeting attention into lasting advocacy and how that fuels the kind of progress that leadership notices and values.

### From Buy-In to Advocacy

Earning stakeholder buy-in is the launchpad. What separates successful projects from stalled ones is continuous connection to business metrics.

- **Link progress to value, not tasks**. Communications should shift from "We closed ten tasks this week" to "Customer onboarding completion rate improved by 15% due to completed optimizations." This is a tactic experts refer to as strategic progress tracking, where the narrative remains tied to business outcomes, helping stakeholders remain engaged with relevance, not utility.
- **Build momentum through meaningful metrics**. Aiming for metrics that matter—like reduced customer churn, improved response time, or incremental revenue—reinforces why the project exists. When stakeholders are

reminded of the underlying purpose every time they receive an update, they remain emotionally and strategically connected.

## Maintaining Strategic Relevance in a Changing Landscape

Projects rarely run in static conditions. Markets shift, leadership priorities pivot, and unforeseen challenges emerge. When those shifts happen, executive attention can wane—unless the project narrative evolves too.

- **Adapt mid-course messaging**. If the strategic landscape changes, your updates should reflect that. For example, if market trends reveal a decline in customer sentiment, reposition the project goal from "feature enhancement" to "improving customer satisfaction by addressing UX gaps." This not only demonstrates awareness but reinforces the project's emerging relevance.
- **Signal early wins**. Every project has points where progress or risk improvements are visible. Share those— "We've already streamlined response time by 10%" or "Risk of system downtime dropped to less than 2%"—to sustain confidence and energy.
- **Use stable narratives during uncertainty**. Iterative delivery builds confidence. In agile or pilot cycles, share completed slices of value rather than waiting for full deliverables. This tactic continually reaffirms the project's progress even when the roadmap shifts.

## Real-World Evidence of Stakeholder Momentum

While specific case examples from research sources can be limited, one useful lens is the empirical study on stakeholder engagement and Lean Six Sigma success in engineering projects.

This research confirms that projects with proactive stakeholder communication, clear alignment of goals, and sustained engagement strategies significantly outperform those lacking these elements[22].

From this, we learn that engaged stakeholders become advocates, not just recipients of updates and that they help projects stay resilient by rallying support when doubts or challenges surface.

In essence, stakeholder engagement is a thread that, when maintained, weaves the project into the fabric of organizational momentum.

Here are actionable moves to convert interest into long-term advocacy:

- **Periodic strategic check-ins**. Instead of just providing status updates, hosting quarterly sessions with stakeholders to discuss business impact and strategic alignment.
- **Ask for advocacy**, not approval. Invite leaders to endorse work: "Would you like to celebrate how this project is improving critical workflows in the next all-hands?"
- **Adjust narrative**, keep consistency. As business goals shift, tweak messaging accordingly—but retain continuity. Stakeholders should feel they're following an evolving story, not jumping between unrelated updates.
- **Share wins**—even small ones. A completed milestone tied to value sends an encouraging message: "We delivered an automation that immediately saved three hours per week for frontline staff."

---

[22] *The Impact of Project Management in Achieving Project Success – Empirical Study*," ResearchGate. https://www.researchgate.net/publication/330480266

Project initiation earns attention; momentum sustains your project's relevance and ensures completion matters. In this chapter, we've seen that stakeholders remain engaged when project storytelling ties to tangible business value, adapts with change, and delivers consistent signals of progress. In doing so, the project evolves from being a standalone effort to becoming a visible force in business strategy.

# Chapter 6 – Earning Your Seat at the Table

In complex organizations, influence is rarely granted outright—it's earned over time through actions, insight, and reliability. Project managers who want a voice in strategic decisions must demonstrate that their work directly advances the organization's goals, and that they understand the pressures executives face and can be trusted to guide resources wisely. This chapter explores how to earn that place at the decision-making table, beginning with the most critical factor of all: trust, the measurable, work-driven trust that persuades leaders to see you as a strategic partner rather than a tactical executor.

## Building Enduring Trust with Executives and Cross-Functional Leaders

In high-stakes projects, particularly those involving multiple departments or global teams, trust is the currency that determines whether your insights are acted upon. Without it, even the most well-supported recommendation can be ignored. Executives operate under constant pressure, balancing stakeholder expectations, market shifts, and budget constraints. When they trust a project manager, they see them as a filter for reliable information, a problem-solver who won't pass along noise, and a steady hand in volatile situations.

Experts in leadership effectiveness and project governance emphasize that trust is built on three pillars: **competence, integrity,** and **relationship alignment**. If any one of these is missing, influence erodes.

## Proven Trust-Building Behaviors

### 1. Demonstrating Competence Through Consistent Delivery

Consistency signals dependability. Executives are more inclined to trust a PM whose projects consistently meet agreed milestones, stay within scope, and adapt gracefully to change. This doesn't mean avoiding difficult conversations, but surfacing potential risks early and presenting realistic solutions.

Practical steps:

- Maintain a track record of clear deliverables backed by evidence-based reporting.
- Use project dashboards that highlight progress as well as how each milestone supports strategic priorities.
- Share lessons learned post-project to show you're committed to improvement.

### 2. Practicing Transparent Communication

Trust is damaged when leaders feel they're only hearing part of the truth. Being transparent means delivering the right information framed for decision-making. Transparency involves acknowledging setbacks as readily as successes. This balance signals professionalism and prevents surprises that undermine credibility.

Practical steps:

- When reporting issues, pair the problem with a proposed resolution or mitigation path.
- Avoid sugarcoating data—leaders value accuracy over temporary comfort.
- Document and share both qualitative feedback and quantitative metrics.

It's tempting to present solutions quickly, but executives may see this as dismissive if you haven't fully understood their perspective. Active listening—summarizing their concerns, asking clarifying questions, and confirming understanding—lays the groundwork for buy-in.

Practical steps:

- Begin strategic discussions by asking leaders about their top three priorities or constraints for the quarter.
- Use that information to shape recommendations in terms they already value.
- Revisit those priorities periodically to show you're aligned with shifting conditions.

## Creating a Shared Language

A common barrier between project managers and executives is language. PMs often speak in terms of deliverables, dependencies, and sprint velocity; executives speak in terms of revenue, market share, and risk exposure. Bridging this gap requires translating project terms into business impact.

A shared language can be built by:

- **Mapping each project** deliverable to an organizational KPI or strategic initiative.
- **Reframing progress updates** to answer: What does this mean for our competitive position?
- **Incorporating the organization's mission, values, and vision** into project narratives so that leadership sees a direct connection.

This alignment improves communication and positions you as someone who gets it from a leadership perspective.

## Navigating Cultural and Functional Differences

In cross-functional or multinational projects, trust-building isn't one-size-fits-all. Organizational cultures vary. Some prioritize directness and speed, while others value consensus and formality. Misreading these cues can unintentionally damage relationships.

Key considerations:

- **Functional culture**: Finance leaders may expect precision in numbers; marketing leaders may value creativity and audience insight. Tailor your approach accordingly.
- **Geographic culture**: In some regions, relationship-building precedes business discussions; in others, efficiency is valued over prolonged rapport-building.
- **Hybrid teams**: Remote or hybrid teams may require extra effort to maintain visibility and rapport, as trust is harder to establish without face-to-face interaction.

Practical approaches:

- **Research functional and regional norms** before initiating strategic discussions.
- **Use diverse communication channels**—video calls for nuance, written summaries for record-keeping.
- **Ensure every stakeholder feels heard** by summarizing input and reflecting it back in decision-making materials.

## Pulling It Together: The Trust-to-Influence Pathway

The journey from being a competent project executor to a trusted strategic voice is incremental:

- **Deliver reliably** – Build a track record executives can point to.
- **Communicate transparently** – Share the real picture, good and bad.
- **Listen actively** – Shape solutions around leadership priorities.
- **Align language** – Speak in terms of business outcomes.
- **Adapt to context** – Recognize and respect cultural and functional differences.

Over time, these behaviors create a reinforcing loop: trust opens doors to higher-level conversations, and participation in those conversations deepens trust.

## Challenging Priorities with Negotiation Skills that Earn Respect

In high-stakes projects, especially those with multiple stakeholders, you will often find yourself in situations where leadership priorities conflict with operational realities. Deadlines may be unrealistic, resources may be insufficient, or proposed changes could derail carefully aligned plans. How you push back in these moments determines whether you are seen as a trusted advisor or a roadblock. This is where negotiation skills become indispensable—grounded in respect, preparation, and a focus on shared objectives.

Tactful pushback is not about saying "no" outright; it requires reframing a directive so leadership understands the risks, trade-offs, and potential alternatives. Instead of focusing on what can't be done, effective project managers highlight why a different approach serves the organization's broader goals.

- **Start with alignment**: Begin by affirming your understanding of the leadership's intent. This prevents the conversation from becoming adversarial.
- **Lead with evidence, not emotion**: Present data, precedents, and outcomes from similar situations rather than personal opinion.
- **Offer a constructive alternative**: Even when disagreeing, propose an option that addresses leadership's concerns while preserving project viability.

For example, if a senior executive insists on an accelerated launch timeline, you could present historical data showing how similar compressed schedules led to post-launch defects, then propose a phased rollout that still meets market demands while reducing risk.

## Negotiation Techniques in Practice

### 1. Preparation & Framing

Drawing from insights from research[23], preparation is the single most important step before entering a priority-challenging conversation.

- **Know your facts**: Have key performance metrics, budget implications, and industry benchmarks ready.
- **Frame in organizational terms**: Instead of saying *"We don't have enough resources,"* say *"Given the current allocation, achieving X goal by Y date may reduce ROI by Z%."*

---

[23] "Effective Project Management Negotiation Techniques." 2024. Awork.com. 2024. https://www.awork.com/glossary/negotiation-techniques-in-project-management.

- **Anticipate objections**: Consider the counterarguments executives might raise and prepare credible responses.

## 2. Interest-Based Negotiation

Popularized in both business and conflict resolution contexts, this approach focuses on interests rather than positions.

- **Interests** are the underlying motivations—such as market share growth, customer retention, or brand reputation.
- **Positions** are the stated demands—such as *"We must launch in Q2."*

By steering the discussion toward shared interests, you avoid zero-sum outcomes. For example, if leadership wants speed and you want quality assurance, explore how a partial feature release could serve both.

## 3. Mutual Gain Solutions

Negotiation is most successful when both parties leave the table with a win-win. This might mean:

- Adjusting scope to fit the timeline without compromising core deliverables.
- Securing additional resources in exchange for accepting a revised milestone.
- Offering a short-term compromise that includes a review period to reassess priorities.

Mutual gain often came from transforming "this or that" scenarios into "this and that" scenarios—by creatively restructuring deliverables.

Even the best-prepared arguments can fail if delivered at the wrong moment or in the wrong tone. Project managers who succeed at influencing decisions pay close attention to both verbal and non-verbal cues:

- **Signs of openness:** Leaning forward, nodding, asking exploratory questions.
- **Signs of resistance**: Crossed arms, clipped responses, checking devices.
- **Pivoting accordingly**: If resistance is high, slow down and ask clarifying questions to surface underlying concerns before making your case.

In cross-functional settings, cultural awareness plays a role here as well. In some cultures, direct disagreement may be frowned upon, requiring more indirect approaches and greater emphasis on relationship-building before challenging priorities.

## Case Insight: Reframing Delay as Risk Mitigation

A research study on negotiation in project management described a scenario where a project manager faced pressure to eliminate a testing phase to meet a launch deadline[24]. Instead of rejecting the directive outright, the PM reframed the delay as a calculated risk mitigation strategy. By showing how skipping the testing phase could expose the company to significant compliance penalties and reputational damage, the PM gained

---

[24] M, Dharshini G., and Sai Yashas G. M. 2020. "A Good Negotiation Process Leads to Success in Project Management." *International Journal of Engineering Research & Technology* 9 (5). https://doi.org/10.17577/IJERTV9IS050835.

executive support for a slightly extended timeline, ultimately avoiding costly post-launch fixes.

In practice, challenging priorities with negotiation skills that earn respect requires you to combine emotional intelligence with strategic acumen. When you can shift the conversation from "my way vs. your way" to "our shared path to the best outcome," you stop being seen as just a project executor and start being recognized as a strategic partner—someone whose insights and judgment help steer the organization toward smarter, more sustainable decisions.

# Leading Strategic Trade-Off Conversations on ROI and Risk

## Moving from Execution to Strategy

Earning a seat at the table requires moving beyond the mindset of "task completer" into the role of "decision partner." This shift is not about abandoning delivery responsibilities, but about complementing them with the ability to influence the direction of projects based on business impact. Strategic trade-off conversations are where this transformation becomes visible. Executives do not expect you to avoid trade-offs but want you to frame them in a way that clarifies options, consequences, and strategic alignment.

The most effective project managers enter these discussions with a full understanding of how the initiative supports the organization's broader goals: market expansion, operational efficiency, innovation leadership, or compliance. By positioning yourself as someone who can bridge project execution with executive decision-making, you build credibility that lasts beyond a single project lifecycle.

## Risk-Return Framing

One of the most valuable contributions you can make in trade-off discussions is reframing decisions in terms executives naturally evaluate: return on investment, cost avoidance, and risk exposure. This means moving away from purely operational language like *"this will take three more weeks"* and toward strategic metrics such as *"this delay would reduce time-to-market by X%, potentially impacting projected revenue by Y%."*

Key elements of risk-return framing include:

- **Quantifying impact on ROI** – Show not only the costs of an option but also the financial upside or protection it provides. For example, a feature delay might reduce immediate spending but cost the company a lost market share.
- **Highlighting cost avoidance** – Present the savings from preventing rework, avoiding regulatory penalties, or mitigating reputational damage.
- **Balancing opportunity and threat** – Acknowledge both potential gains and risks, giving executives a complete view for decision-making.

Here's a simple structure you can use:

| Option | Potential Benefit | Potential Risk | Net Business Impact |
|---|---|---|---|
| Implement now | Faster customer adoption, brand leadership | Higher defect rate, support costs | +8% projected market share but +5% service costs |
| Delay for QA | Higher quality, fewer customer complaints | Lost first-mover advantage | Neutral market share, stronger brand reliability |

Translating technical trade-offs into executive metrics like revenue growth, customer satisfaction scores, or regulatory compliance, you help leaders make decisions with clarity and confidence.

## Facilitating Consensus Under Pressure

Trade-off conversations often become emotionally charged when deadlines loom or competitive pressures intensify. In such moments, the role of a strategic PM is to keep the discussion anchored in facts and framed around the business's strategic objectives.

Practical techniques include:

- **Set shared decision criteria early** – Before discussing solutions, agree on what will guide the decision (e.g., ROI threshold, risk tolerance, time-to-market goals).
- **Use scenario modeling** – Show best-case, worst-case, and most-likely outcomes for each option. This helps depersonalize the conversation and makes it harder for subjective bias to dominate.
- **Reference precedent** – Where possible, bring in examples of similar past decisions and their outcomes, whether internal or from industry benchmarks.
- **Manage participation dynamics** – Ensure quieter voices are heard, especially from functional areas directly affected by the trade-off. Their input often surfaces hidden risks or overlooked opportunities.

A disciplined approach to facilitating consensus also means knowing when to pause a decision until more data is available. It's better to delay a choice briefly to avoid long-term misalignment than to rush toward a commitment that will erode trust later.

## Maintaining Momentum Post-Decision

Your influence doesn't end when the trade-off is agreed upon. Executives and stakeholders are far more likely to trust your recommendations in future decisions if you consistently demonstrate that past decisions have delivered the intended results. Post-decision actions that reinforce strategic value include:

- **Link progress updates to original business objectives** – Instead of generic status reports, tie achievements back to the ROI, risk mitigation, or compliance goals agreed upon during the trade-off conversation.
- **Track unintended consequences** – If a decision produces side effects, positive or negative, document them. Sharing this learning shows transparency and sharpens future trade-off evaluations.
- **Reinforce the rationale** – Periodically restate why a particular path was chosen, especially if internal stakeholders change. This preserves institutional memory and prevents second-guessing.

The ability to lead trade-off conversations that weigh ROI and risk is one of the clearest markers that you have earned your seat at the table. It demonstrates that you are not just delivering against a plan but actively shaping decisions that define the organization's trajectory.

By consistently framing discussions in business terms, keeping them fact-based under pressure, and showing measurable follow-through, you evolve from project executor to strategic partner. The reward is a stronger voice in high-stakes conversations and also a deeper trust from executives who see you as integral to navigating complex choices in a competitive environment.

# Part III

# The Skills: Business Acumen Every PM Must Master

# Chapter 7 – Financial Fluency for PMs

Project managers often focus on schedules, scope, and stakeholder expectations, but the reality is that no project exists in isolation from the organization's financial objectives. Every decision, from selecting vendors to adjusting timelines, has a ripple effect on profitability, cash flow, and long-term strategic positioning. Yet many PMs treat finance as a separate function, relying solely on accountants or controllers to handle the numbers. This creates a dangerous blind spot.

Financial fluency empowers PMs to interpret financial data, connect it to project outcomes, and communicate in the language executives trust. It requires the financial drivers that determine whether a project is truly successful. This chapter will explore how to translate financial concepts into actionable project decisions, spot early warning signs of trouble, and align deliverables with the metrics that matter most in the boardroom.

## The Strategic Lens: Linking Projects to Business Objectives

### Why Strategic Alignment Matters

An organization's projects are investments, not isolated tasks. Each project competes for limited resources and must justify its existence by contributing to defined business objectives. When PMs understand how their projects fit into revenue growth, cost control, market expansion, or risk reduction, they can prioritize effectively and defend decisions in executive discussions.

Research emphasizes that organizations with mature strategic alignment practices achieve far higher project success rates because PMs make decisions in the context of business strategy, not just task completion[25]. Strategic alignment means more than understanding the business case. It requires knowing which financial levers your project affects, whether that's increasing gross margin, reducing operational costs, or protecting cash flow during a downturn.

## Translating Project Outcomes into Financial KPIs

Executives assess projects through metrics such as:

- **Gross Margin:** The percentage of revenue left after direct costs are subtracted. For PMs, this can be influenced by controlling supplier costs, optimizing labor efficiency, or reducing rework.
- **EBITDA:** Earnings before interest, taxes, depreciation, and amortization. This is a measure of operational profitability, and project delays or inefficiencies can directly lower it.
- **Cash Flow:** The timing of cash inflows and outflows. Payment schedules, milestone billing, and procurement strategies all affect this.

For example, if a software rollout delays customer onboarding by two months, the impact is on delivery as well as on deferred revenue recognition, which could hurt quarterly financial results. Understanding this link enables a PM to escalate appropriately and propose financially viable solutions.

---

[25] Cohen, Dennis. 2000. "Why Finance Matters for Project Managers - Important Component." Www.pmi.org. 2000.
https://www.pmi.org/learning/library/finance-matters-project-managers-important-1861.

## Financial Terms PMs Should Master

While basic budget tracking is essential, strategic fluency comes from understanding deeper concepts:

| Term | Why It Matters for PMs |
|---|---|
| Break-even Analysis | Identifies the point where project revenues match costs. Helps assess feasibility and timeline sensitivity. |
| Contribution Margin | Revenue minus variable costs. Shows how much a project contributes to covering fixed costs and generating profit. |
| Gross Margin | Indicates profitability after direct costs. PMs can improve it through efficiency and cost control. |
| Working Capital | The capital available for day-to-day operations. Delayed deliverables or invoicing can tie up capital. |
| Accrual vs. Cash Accounting | Understanding revenue recognition and expense recording helps in reporting and forecasting accurately. |

For example, in a capital-intensive infrastructure project, extending the delivery phase without milestone billing can erode working capital, putting pressure on the entire business.

## Bridging the Gap Between Finance and Delivery

Finance teams often focus on long-term returns, while PMs manage near-term execution. The bridge between the two is clear, quantifiable communication. This includes:

- Mapping project milestones to revenue or cost impact dates.
- Explaining change requests in terms of ROI and payback period, not just "scope impact."

- Using consistent terminology with finance to avoid misinterpretations.

Research highlights that leaders respond better to proposals framed in financial outcomes rather than operational outputs[26]. For example, instead of saying, *"We need two more weeks to complete testing,"* a PM might say, *"Completing testing now reduces the risk of a post-launch defect, which could save us an estimated $250,000 in warranty claims."*

## Spotting Early Financial Red Flags

Financial distress in a project rarely happens overnight—it builds through patterns:

- **Budget Burn Rate Exceeding Plan**: Indicates overspending that may require scope re-evaluation.
- **Margin Erosion**: Caused by rising costs or underestimated work effort.
- **Revenue Recognition Delays**: Especially critical in milestone-based contracts.
- **Supplier Payment Compression**: May signal cash flow issues upstream in the organization.

PMs who identify these signs early can recommend corrective actions such as renegotiating vendor contracts, re-phasing deliverables, or accelerating revenue-generating components.

---

[26] Landry, Lauren. 2018. "Financial Terminology: 20 Financial Terms to Know | HBS Online." Business Insights - Blog. October 11, 2018. https://online.hbs.edu/blog/post/finance-for-non-finance-professionals-finance-terms-to-know.

When PMs master financial fluency, they stop being viewed as operational managers and start being recognized as strategic contributors. This shift influences:

- **Decision-Making Authority:** PMs with financial insight are trusted with larger budgets and higher-stakes initiatives.
- **Cross-Functional Respect:** Finance and executive teams view them as partners, not just requesters.
- **Career Progression**: Strategic PMs often transition into portfolio management or executive leadership roles.

In practice, this means that understanding financial KPIs is a competitive advantage that ensures your project's success is both operationally and financially sustainable.

## Reading Between the Lines of a P&L

Most project managers see a Profit & Loss (P&L) statement as the finance department's territory. But learning to interpret it unlocks a wealth of insights into project efficiency, cost risks, and hidden opportunities that typical project dashboards never reveal. A P&L tells the story of how organizational resources translate into profit or losses and where a project's financial footprint fits into that story.

### Dissecting the P&L for Project Insight

At its core, a P&L summarizes revenue, costs, and profit over a period. But for PMs, the real value lies in breaking down specific line items:

- **Revenue Impact**: Projects tied to product launches or service upgrades influence top-line growth. If a delay shifts revenue recognition into the next quarter, leadership needs that visibility early.

- **Cost of Goods Sold (COGS):** This includes raw materials, vendor costs, or direct labor. Projects affecting manufacturing efficiency or procurement contracts can directly lower COGS.
- **Operating Expenses:** Overheads like IT systems, facilities, or shared services often hide cost leakages. For instance, a project prolonging its timeline might drive up administrative costs without anyone noticing.
- **Depreciation and Amortization:** Capital-intensive projects, such as data centers or machinery upgrades, impact these non-cash expenses. PMs who track asset utilization can prevent underperforming investments from eroding profitability.

By linking these numbers back to specific project activities, PMs shift the conversation from "Is the project on track?" to "Is the project financially worthwhile?"

## Spotting Hidden Opportunities and Risks

A close read of P&L trends often uncovers signals for both risks and opportunities:

- **Escalating Vendor Costs:** If COGS creeps upward, it might be time to renegotiate supplier terms or explore alternative sourcing.
- **Underutilized Assets:** High depreciation with low corresponding revenue could flag projects that need repurposing or divestment.
- **Scope for Upsell:** A marketing automation project reducing customer acquisition cost might justify expanding the scope to cross-sell initiatives.

Finance teams may focus on totals, but PMs can interpret fluctuations at the project level, enabling proactive interventions before small inefficiencies become systemic problems.

## Opportunity Cost in Real Terms

Every delayed milestone or misallocated resource carries a price tag beyond the project budget. Opportunity cost reframes these setbacks in terms that executives understand:

- **Lost Revenue**: A two-month delay in launching a subscription service might mean missing an entire billing cycle, reducing annual recurring revenue targets.
- **Reduced ROI:** Extending timelines without increasing benefits dilutes overall returns. A project expected to generate a 25% ROI may drop to 15% if delivery drags six months longer.
- **Risk Exposure**: Cybersecurity initiatives delayed past regulatory deadlines can lead to fines or reputational damage far exceeding the project cost.

When PMs present trade-offs in opportunity cost terms, they transform abstract schedule slippages into concrete business consequences, strengthening their voice in executive decisions.

## Making Financial Analysis Actionable for PMs

To integrate P&L insights into project oversight, PMs can:

- Collaborate with finance early to map project milestones against financial reporting periods.
- Use variance analysis to explain deviations in cost or revenue impact at the project level.
- Tie change requests to updated ROI projections rather than just revised timelines or budgets.

This approach reframes financial analysis from a passive reporting exercise into a strategic decision-making tool where PMs link day-to-day execution to the organization's bigger financial picture.

## Crafting Unshakable Business Cases

Project managers often underestimate how much weight a well-crafted business case carries in executive decision-making. A polished project plan may earn polite nods, but only a business case tied directly to strategic outcomes secures funding and organizational commitment. To influence at this level, PMs need to think like investment analysts, structuring arguments around value, risk, and return rather than tasks, timelines, or technology upgrades.

### Building Executive-Ready Proposals

Executives operate in a world of trade-offs—every dollar, hour, and resource allocated to one initiative comes at the expense of another. That's why business cases need to go beyond technical feasibility to answer three boardroom questions:

- **Return on Investment (ROI):** What measurable value will this project create, and when?
- **Payback Period:** How quickly will the initial investment be recovered?
- **Total Cost of Ownership (TCO):** What are the full lifecycle costs, including implementation, training, support, and eventual replacement or decommissioning?

Framing proposals around these metrics changes the conversation. Instead of selling a new CRM system because "the old one is outdated," a PM might argue that "a 20% improvement in sales conversion from the new system is projected to generate $3M in incremental annual revenue, achieving payback in 18 months."

## The Financial Case for Saying "No"

Strategic PMs also recognize when not to proceed. Many organizations waste resources on initiatives with weak business alignment simply because no one presented the downside in hard numbers. A robust business case includes:

- **Alternative Scenarios**: Best case, expected case, and worst case outcomes for ROI and timeline.
- **Risk Exposure:** Quantified probabilities for cost overruns, delays, or market changes that could erode benefits.
- **Opportunity Cost:** What other strategic investments must be delayed or abandoned if this project proceeds?

By framing risks in business language, PMs give leadership permission to pivot early rather than cancel projects after millions are spent with little return.

## The Business Case Checklist for PMs

To ensure consistency, PMs can use a structured checklist:

| Step | Focus Area | Key Questions to Answer |
|---|---|---|
| 1 | Define the Problem | What measurable pain point or missed opportunity does this project address? |
| 2 | Quantify Expected Benefits | What revenue gains, cost savings, or risk reductions will result? Use conservative estimates. |
| 3 | Analyze ROI Sensitivity | How do outcomes change if costs rise by 10% or benefits fall by 15%? |

| Step | Focus Area | Key Questions to Answer |
|------|-----------|------------------------|
| 4 | Map Risk Mitigation Costs | What contingency plans exist, and what will they cost if triggered? |
| 5 | Link to Strategy & Compliance | How does this align with corporate objectives or regulatory mandates? |

Following this structure transforms business cases from generic requests into boardroom-ready decision tools.

## Making the Business Case Persuasive

Numbers alone rarely secure approval; executives need context. PMs can strengthen their argument by:

- **Linking Benefits to KPIs Executives Track:** Revenue growth, market share, customer satisfaction, or regulatory compliance.
- **Including Comparative Benchmarks:** Industry ROI ranges or competitive performance data help position the project within a broader business landscape.
- **Outlining Non-Financial Gains:** Employee productivity, brand reputation, or sustainability outcomes often carry strategic weight beyond pure dollars.

Financial fluency transforms PMs from cost reporters into value creators. By interpreting financial statements, calculating ROI trade-offs, and presenting scenarios with strategic clarity, PMs elevate their role from project execution to enterprise leadership. The ability to craft business cases that withstand scrutiny ensures projects compete—and win—for scarce resources in the executive arena.

# Chapter 8 – Strategic Prioritization That Drives Impact

In today's organizations, competing initiatives, shrinking budgets, and rapid market shifts create constant noise. Leaders often assume that traditional tools—like urgency/importance matrices—will bring clarity. Yet, as projects multiply across portfolios and decisions span functions and geographies, these "classic" approaches falter. They oversimplify trade-offs, ignore interdependencies, and fail to connect priorities with strategic outcomes. This chapter examines why legacy methods break down at scale, and what modern frameworks offer instead. By understanding these shortcomings, project managers can move beyond firefighting to help organizations allocate time, capital, and talent where they drive real impact.

## Why "Classic" Matrices Break at Scale

### Task-Level Tools vs. Portfolio Complexity

Classic tools like the Eisenhower Matrix were never designed for today's enterprise environments. Originally conceived for personal productivity, they work well when a single decision-maker sorts tasks into "urgent/important" quadrants to manage their time. At a portfolio level, however, the reality looks nothing like a to-do list.

Consider a global product launch with marketing, engineering, finance, and operations involved. A single delay in regulatory approvals might stall an entire release despite dozens of other urgent tasks completed on time. Monday.com notes that while action-priority grids improve individual focus, they ignore portfolio variables such as resource capacity, cross-project dependencies, and market timing windows. The result? Teams optimize for activity rather than outcomes, delivering outputs while strategic goals remain unmet.

Portfolio decisions require visibility into capital allocation, risk exposure, and revenue impact across multiple workstreams—factors no simple 2×2 diagram can accommodate.

*Example— Dependency shock*

A global launch clears 90% of tasks, yet a two-week regulatory slip in one region blocks release everywhere. Portfolio impact: an 8-week cost of delay at:

$$\sim\$150k/week = \sim\$1.2M$$

The matrix looked "green," the portfolio result did not.

## Hidden Variables That Change Everything

As organizations scale, prioritization must account for dimensions beyond urgency and importance. Research highlights several overlooked variables[27]:

- **Value Decay Over Time**: Some initiatives lose potential revenue or competitive advantage if delayed. A feature launch six months late might miss an entire buying cycle, slashing ROI regardless of original "importance."
- **Cost of Delay**: Lean product management often quantifies this explicitly—showing the financial loss per week of postponement. Classic matrices offer no mechanism for this calculation.

---

[27] Levay, Peter. 2024. "13 Product Prioritization Frameworks & How to Pick the Right One." As We May Think — Products & Tools for Thought. May 27, 2024. https://fibery.io/blog/product-management/best-product-prioritization-frameworks/.

- **Regulatory and Compliance Deadlines**: In sectors like finance or healthcare, missing a compliance date carries penalties that outweigh subjective "importance" labels.
- **Customer Segmentation:** A fix affecting 60% of the user base carries different weight than one impacting 5%, even if both seem "urgent."

By failing to model these variables, traditional tools encourage decisions based on gut feeling or loudest-voice bias rather than economic or strategic value.

*Example: Cost of delay*

If the Q4 promo campaign shifts to Q1, expected incremental revenue drops from $2.4M to $1.6M due to seasonality. Cost of delay for 8 weeks ≈ $800k.

*Which framework surfaces this faster—Impact/Effort or WSJF?*

## Governance and the Missing Decision Infrastructure

Portfolio prioritization rarely ends with a single person's quadrant exercise. Board-level or C-suite decisions demand structured debate, scenario modeling, and documented rationale for accountability.

Research emphasizes that effective decision-making at senior levels relies on:

- Pre-reads summarizing options, assumptions, and trade-offs.
- Dialog structures ensuring all voices—finance, risk, operations—are heard before commitment.
- Decision logs capturing not just the choice but why it was made, enabling future review.

Classic matrices lack this infrastructure. They produce rankings without offering transparency into assumptions or dissenting opinions. As stakes rise—multi-million-dollar capital projects, product sunsets, market entries—this absence of rigor erodes trust and leads to constant revisiting of final decisions.

*Try this — Decision log (example):*

| Decision | Options considered | Rationale | Risks noted | Review date |
|---|---|---|---|---|
| Fund APAC launch for Q3 | Delay to Q4, Phase rollout by region, or Full fund for Q3 launch. | ROI projections exceeded 22%; time-to-value within 90 days; aligns with growth targets. | Supply chain vendor dependency; potential regulatory review delays in two regions. | Q3 Monthly Business Review (MBR). |

## Priority Inflation: When Everything Becomes P1

Experts warn of a common failure mode: when simple frameworks meet organizational politics, everything becomes "top priority." Without explicit tiering rules or capacity limits, leaders label too many initiatives as critical, creating gridlock and burnout.

Robust prioritization systems prevent this through:

- **Defined tiers** (e.g., P1 for regulatory/legal, P2 for strategic growth, P3 for incremental improvement).
- **Capacity-linked gates:** Limiting how many P1 items can exist per quarter based on available resources.
- **Sunset clauses**: Requiring periodic re-validation so stale priorities don't clog the pipeline.

Classic tools rarely enforce these disciplines. They present static snapshots, while real organizations need dynamic systems balancing ambition with capacity constraints[28].

*Guardrail — Tiering & capacity caps:*

- Max three P1 initiatives per quarter per portfolio.
- Expedite class requires CFO/COO sign-off.
- P1s must include cost-of-delay estimate.
- Auto-sunset any priority label after 90 days unless re-validated.

## The Illusion of Productivity

Perhaps the biggest danger is psychological. Teams using simple urgency/importance tools often feel productive—lots of tasks completed, colorful matrices filled, meetings held. Yet as Product School notes, this "busy equals impact" illusion hides the fact that many completed initiatives show negligible ROI or strategic contribution.

Without connecting priorities to business metrics like revenue growth, cost savings, or customer retention, organizations risk celebrating motion over progress. Classic tools, by ignoring outcome tracking, let this illusion persist unchallenged.

Acknowledging these gaps sets the stage for modern frameworks—RICE, MoSCoW, Cost of Delay, WSJF—that incorporate economic impact, uncertainty, and capacity planning. Unlike Eisenhower grids, these methods scale across

---

[28] Peterka, Peter. 2024. "COVID-19 Situation: Six Sigma Ongoing Training Announcements." SixSigma.us. September 12, 2024. https://www.6sigma.us/project-management/levels-of-priority/.

portfolios, support governance needs, and connect execution choices to enterprise strategy.

## Choosing the Right Framework for the Job

Selecting the right prioritization framework isn't just a matter of preference—it determines whether decisions create strategic clarity or degenerates into political debate. Each framework carries built-in assumptions about data maturity, decision scope, and time sensitivity. Below, we map widely used models to the problems they solve best, outline their limitations, and show where organizations often misuse them.

A practical takeaway is this: no single framework wins everywhere. A mature portfolio might blend two or three methods, each addressing different layers of decision-making—product backlog grooming, quarterly portfolio planning, or executive capital allocation.

### 1. Impact/Effort & Value Scoring: Quick Wins, Limited Depth

**Where it works:** Small teams, early-stage startups, or quarterly "housekeeping" prioritization where speed beats precision. Ideal for backlog grooming when initiatives are bite-sized and data is scarce.

**How it works:** Ideas get plotted by impact (value created) and effort (cost or complexity). High-impact/low-effort ideas naturally bubble up as "quick wins."

**Strengths:**

- Simplicity: No financial modeling required; anyone can facilitate.
- Engagement: Teams visualize trade-offs instantly, reducing debate fatigue (TeamGantt; ProductLift).

**Weaknesses:**

- **Lacks time sensitivity**: No way to model cost of delay or value decay.
- **Subjectivity risk:** Impact often devolves into gut feel without supporting metrics.
- **Poor scalability**: Breaks down when 50+ initiatives require cross-functional coordination.

**Use when:** Cleaning up product backlogs, prioritizing UX tweaks, or running design sprints—not when millions in revenue ride on timing.

## 2. RICE (Reach, Impact, Confidence, Effort): Balancing Ambition and Risk

**Where it works:** Product teams managing features, growth experiments, or market-facing roadmaps where reach and confidence matter.

**Formula:**

$$Score = \frac{Reach \ x \ Impact \ x \ Confidence}{Effort}$$

**Strengths:**

- **Quantifies uncertainty**: Confidence factor forces teams to rate evidence strength.
- **Market-aware**: Reach ground-level decisions in potential customer exposure.
- **Relative scoring**: Compares options systematically rather than emotionally.

**Weaknesses:**

- **Impact inflation:** Teams overestimate benefits without historical baselines.

- **Data demands:** Works poorly if market or effort estimates are immature.
- **Interdependency gaps:** Doesn't model portfolio-level constraints.

**Use when:** Product backlogs need structured scoring; leadership wants semi-quantitative prioritization but can tolerate rough confidence bands.

### 3. ICE (Impact, Confidence, Ease): Speed Over Precision

**Where it works:** Growth teams running rapid A/B tests, marketing experiments, or MVP iterations.

**Formula:**

$$Score = \frac{Impact \times Confidence \times Ease}{10}$$

**Strengths:**

- **Fast cycles:** Scoring takes minutes, keeping experimentation velocity high.
- **Low cognitive load:** Fewer variables than RICE; easy adoption by non-technical teams.

**Weaknesses:**

- **Oversimplification:** Ignores revenue modeling, customer segmentation, or risk exposure.
- **Short-term bias:** Rewards ease too heavily, underplaying strategic initiatives with long payback periods.

**Use when:** Marketing or growth squads need weekly prioritization without complex financial inputs. Avoid using it for portfolio investment decisions.

## 4. MoSCoW: Release Planning Workhorse

**Where it works:** Agile teams scoping MVPs, release cycles, or quarterly planning workshops.

**Buckets:**

- **Must-have:** Non-negotiable requirements for release viability.
- **Should-have:** Important but negotiable if capacity shrinks.
- **Could-have:** Nice-to-have features.
- **Won't-have:** Explicitly deprioritized items for this cycle.

**Strengths:**

- **Stakeholder alignment**: Simple language bridges tech and business audiences.
- **Scope control:** Prevents "everything is critical" syndrome if enforced properly.

**Weaknesses:**

- **Must-have creep**: Without ranking inside buckets, 80% of items land in "Must-have," killing focus.
- **No economic framing**: Buckets lack ROI or cost modeling; purely categorical.

**Bucket ranking rule**: Within "Must-have," rank by Value Score or RICE; cap Must-haves at 50% of capacity.

**Use when:** Facilitating MVP discussions or sprint scope alignment, but layer with scoring methods for economic prioritization.

## 5. Kano / Opportunity Scoring: Voice-of-Customer Anchors

**Where it works:** Customer-centric roadmaps where delight, satisfaction, or competitive differentiation matter (Product School; Fibery).

**Features are classified as:**

- **Basic expectations:** Missing them causes dissatisfaction but meeting them doesn't delight.
- **Performance attributes**: More investment → higher satisfaction.
- **Delighters:** Unexpected features creating customer loyalty spikes.

**Strengths:**

- **Customer-backed**: Ties prioritization to real satisfaction data, not internal opinions.
- **Strategic edge:** Identifies features creating market buzz vs. parity maintenance.

**Weaknesses:**

- **Research-heavy**: Requires surveys, interviews, or NPS analytics.
- **Indirect ROI links:** Customer delight ≠ immediate revenue unless paired with financial modeling.

**Use when**: Differentiation or customer retention drives strategy; combine with RICE or Cost of Delay for economic rigor.

## 6. Cost of Delay / WSJF: Economics Meets Flow

**Where it works:** Scaled Agile environments, portfolio Kanban, or any context with queues, bottlenecks, and high time sensitivity.

**Formula (Weighted Shortest Job First):**

$$WSJF = \frac{Cost\ of\ Delay}{Job\ Duration}$$

**Strengths:**

- **Explicit economics**: Shows dollars lost per week of delay; aligns with Lean Portfolio Management.
- **Queue discipline**: Prioritizes high-value, short-duration items to maximize flow efficiency.
- **Capacity alignment**: Balances value delivery with resource constraints.

**Weaknesses:**

- **Estimation challenges**: Cost of Delay often inferred rather than measured.
- **Change resistance**: Finance or legacy PMOs may resist economic language replacing traditional scoring.

**Use when:** Time-critical portfolios face regulatory deadlines, competitive launches, or resource bottlenecks.

## 7. Priority Levels & Service Classes: Guardrails Against Chaos

**Where it works:** Operational environments needing service-level agreements (SLAs) and explicit escalation paths.

**Structure:**

- **P1:** Regulatory/legal or revenue-critical.
- **P2:** Strategic growth initiatives.
- **P3:** Incremental or exploratory work.
- **Expedite class:** Overrides normal queues under predefined conditions.

**Strengths:**

- **Clarity:** Removes ambiguity during firefighting scenarios.
- **Governance fit**: Links priority classes to funding or capacity allocation policies.

**Weaknesses:**

- **Static risk:** Lacks dynamic scoring; best paired with methods like WSJF or RICE.

**Use when:** Organizations suffer from "everything is urgent" culture; combine with economic models for final sequencing.

## Framework Selector Cheat Sheet

| Decision Context | Best Frameworks | Why | Avoid |
|---|---|---|---|
| Backlog grooming, UX tweaks | Impact/Effort, ICE | Fast, lightweight, facilitation-friendly | WSJF (overkill for small items) |
| Product feature roadmaps | RICE, Kano + RICE combo | Balances customer reach with economic impact | MoSCoW alone (lacks ROI rigor) |
| MVP scoping, sprint planning | MoSCoW + Value Scoring | Aligns stakeholders, controls scope creep | ICE (too shallow for releases) |
| Regulatory deadlines, time-critical | WSJF, Cost of Delay | Quantifies economic loss of delay, aligns with flow | Impact/Effort (no time factor) |
| Portfolio capital allocation | WSJF + Priority Levels | Links investment, capacity, and strategic metrics | Kano alone (customer delight bias) |

### Avoiding Anti-Patterns

- **Over-Scoring Syndrome**: Teams spend weeks perfecting scores while market windows close.
- **Single-Framework Dogma**: Mature portfolios mix methods—e.g., Kano for voice-of-customer, WSJF for sequencing, MoSCoW for release scoping.
- **Ignoring Confidence Levels**: Impact scores without confidence ratings fuel political gaming.

# From Backlog to Boardroom – Operationalizing Prioritization

The real challenge in prioritization is embedding a framework into organizational rhythms so that decisions are repeatable, transparent, and trusted across teams and leadership levels. A scoring model on paper means little if meetings devolve into politics or decisions vanish into PowerPoint decks without follow-up. Operationalizing prioritization bridges the gap between theory and execution, ensuring that whether you're grooming a backlog at the squad level or debating trade-offs in the boardroom, decisions follow a structured, evidence-driven process rather than gut feel or hierarchy.

Below, we break down the building blocks of a robust prioritization operating system.

### Upstream Hygiene: Preparing Decision-Ready Items

Poor prioritization often starts with poor inputs. Teams show up to planning sessions with vague ideas—"improve user experience," "enhance analytics"—leaving executives to debate assumptions rather than choices. Adobe for Business emphasizes the need for backlog hygiene so each item entering the prioritization funnel contains minimum decision data:

- **Hypothesized Business Value**: Revenue growth, cost reduction, risk mitigation, or strategic alignment explained in one or two lines.
- **Effort Ranges**: T-shirt sizing or story points converted into approximate cost or timeline ranges.
- **Dependencies:** Noting upstream or downstream initiatives avoids approving items that later hit hidden bottlenecks.
- **Risk Notes:** Regulatory deadlines, operational fragility, or reputational exposure if delayed.
- **Definition of Ready (DoR):** A checklist ensuring no item moves forward without evidence or owner clarity.

*Example – Poor vs. Prepared Backlog Item*

| Criteria | Poorly Prepared Item | Decision-Ready Item |
|---|---|---|
| Description | "Improve analytics" | "Add real-time dashboards for top 3 KPIs tied to $2M ARR accounts" |
| Effort Estimate | "Medium" | "8-week dev effort; dependent on API integration project" |
| Business Impact | "Better UX" | "Expected 15% reduction in churn → $600K retention impact" |
| Risk/Deadline Notes | None | "Regulatory compliance deadline in Q3" |
| Owner | Unassigned | Product Analytics Lead assigned |

By standardizing inputs, meetings focus on trade-offs, not translations, accelerating decision cycles and reducing rework later.

## Structured Criteria and Decision Cadence

Even with clean inputs, prioritization collapses without a shared rubric and regular cadence. Experts highlight three pillars:

- **Published Criteria**: Weightings for value, risk, confidence, or cost should be documented, versioned, and visible to all contributors. Hidden formulas fuel political debate; transparency builds trust.
- **Decision Logs:** Each cycle should record the rationale, dissenting views, and assumptions behind final rankings. When priorities are revisited next quarter, leaders see why a choice was made rather than repeating old arguments.
- Cadenced Reviews:
  - Monthly portfolio reviews for incremental reprioritization.
  - Quarterly strategy resets aligning with financial planning and KPI cycles.
  - Ad-hoc escalation paths for regulatory or crisis-driven items.

*Sample Prioritization Cadence*

This rhythm prevents "fire-drill prioritization," where every meeting feels like starting from scratch.

| Cadence | Purpose | Participants | Output |
|---|---|---|---|
| Weekly Grooming | Prep backlog items for scoring | Squad leads, Product Owners | Clean, decision-ready backlog |
| Monthly Review | Apply scoring + adjust sequencing | Cross-functional PMs, Finance | Updated portfolio priorities |
| Quarterly Reset | Align with KPIs, budgets | Execs, Portfolio Steering Comm | Strategic roadmap adjustments |
| Ad-hoc Escalation | Handle urgent regulatory changes | Risk, Ops, PMO leads | Fast-tracked approvals or pivots |

## Cross-Functional Facilitation: Reducing Bias and HiPPO Sway

Prioritization often derails when the HiPPO—Highest Paid Person's Opinion—dominates despite limited context. The recommended structured facilitation practices include:

- **Pre-reads:** A single-slide summary per initiative, circulated 48 hours before meetings, forces preparation and reduces real-time information overload.
- **Scenario Tables:** Showing multiple scoring outcomes—e.g., financial weighting vs. customer-impact weighting—surfaces trade-offs transparently.
- **Cross-Functional Voices:** Finance validates ROI math, Risk assesses compliance exposure, CX teams weigh customer experience impact.

## Facilitation Techniques That Work

- **Time-boxed discussions:** 10 minutes per item prevents endless debates.
- **"Silent scoring" before open discussion**: Reduces anchoring bias from senior voices.
- **Rotating facilitators**: Prevents process ownership from defaulting to one department (often PMO).

This ritualization of fairness builds credibility so even deprioritized teams see the process as legitimate rather than political.

## Guardrails and Anti-Patterns

Even mature systems fail without boundaries. Common traps include score gaming, urgency inflation, and stale backlogs. Insights from research suggest explicit safeguards:

| Anti-Pattern | Consequence | Guardrail Solution |
|---|---|---|
| Score Gaming | Teams inflate value or deflate effort | Require evidence links + confidence ratings on all scores |
| "Everything is Urgent" Culture | Resource burnout; no sequencing discipline | Capacity caps: Max P1 items per quarter; service classes (Expedite, Standard, Deferred) |
| Stale Priorities | Old items clog backlog, losing relevance | Auto-sunset scores older than one quarter unless revalidated |
| Endless Re-Scoring Cycles | Decision paralysis | Lock scores post-decision; revisit only at next cadence window |

These rules transform prioritization from open-ended debate into a time-bound, evidence-backed process.

## Closing the Loop: Communicating and Learning

A decision ignored after the meeting is as bad as no decision at all. Here's the importance of post-decision discipline:

- **Roadmap Updates**: Within 48 hours, publish revised priorities in portfolio tools (e.g., Jira, Aha!, Monday.com).
- **Trade-Off Communication:** Explain what dropped as clearly as what advanced; transparency prevents shadow lobbying later.
- **Learning Feedback**: After major launches, feed outcome data—ROI achieved, delays faced—back into scoring templates. Over time, assumptions get sharper, reducing estimation bias.

- Decision meeting ends → Portfolio PM updates roadmap by EOD.
- Comms lead sends summary email: "3 items advanced, 2 deferred, rationale linked."
- Next quarterly review compares forecasted vs. actual impact for top 5 items.

This feedback loop turns prioritization into a learning system, not just a gating mechanism.

When all elements—clean inputs, published criteria, regular cadence, cross-functional facilitation, guardrails, and closing loops—work together, organizations move from ad-hoc debates to a prioritization operating system.

Key outcomes include:

- **Speed:** Decisions made in hours, not weeks.
- **Trust:** Even deprioritized teams accept results as fair.
- **Alignment:** Roadmaps reflect strategy, not politics.
- **Learning:** Each cycle improves data quality and estimation accuracy.

## Measuring Decision Quality, Not Just Throughput

Fast decision-making means little if outcomes fail to deliver value. Organizations need post-decision metrics that separate speed from effectiveness. Tracking the percentage of prioritized items hitting target KPIs shows whether frameworks translate into real business impact. Comparing forecasted vs. actual ROI uncovers over-optimistic scoring or missed assumptions. Metrics such as average time to decision, percentage of expedited items, and rework due to unclear criteria highlight process inefficiencies without waiting for annual reviews.

Equally important are retrospectives on choices. A quarterly review of "best calls" and "worst calls" exposes blind spots— initiatives deprioritized but later revived, or rushed items that failed to deliver. Rather than adding more meetings, teams refine scoring rubrics, confidence ranges, and decision thresholds based on these insights. Over time, this feedback loop evolves prioritization into a learning system where every cycle improves both decision speed and strategic accuracy, not just task throughput.

## Practice Lab — Strategic Prioritization

### Exercise 1 — Detect cost of delay (10 minutes)

You have three initiatives with seasonal revenue. Estimate cost of an 8-week delay for each and rank by loss per week.

- **A**: Peak window now, $1.6M expected, drops to $1.0M next quarter.
- **B:** Stable $900k any quarter.
- **C**: $2.2M if launched before holiday; $1.2M after.

**Deliverable**: Cost of delay per week and a short rationale.

### Exercise 2 — Score a mini backlog (15 minutes)

Compute RICE for five items *(Reach/mo, Impact 0.25–3, Confidence %, Effort person-weeks)*. Capacity this quarter is 20 person-weeks. Select a portfolio.

- **Onboarding revamp**: R=3,000, I=1, C=70%, E=10
- **Pricing test**: R=800, I=2, C=80%, E=3
- **Mobile perf fix:** R=5,000, I=0.5, C=90%, E=2
- **New referral flow**: R=1,200, I=1.5, C=60%, E=6

- **Fraud rules v2:** R=700, I=3, C=50%, E=8

**Deliverable:** Scores, chosen set within 20 weeks, one trade-off note.

## Exercise 3 — WSJF sequencing (10 minutes)

Compute WSJF and order work.
- **D:** CoD $90k/wk, Duration 2 wks
- **E:** CoD $60k/wk, Duration 1 wk
- **F:** CoD $140k/wk, Duration 5 wks

**Deliverable:** Sequence and one sentence why.

## Exercise 4 — Build your decision cadence (10 minutes)

Fill the grid for your context.

| Cadence | Purpose | Who attends | Key artifacts | SLAs |
|---|---|---|---|---|
| Weekly grooming | | | | |
| Monthly portfolio | | | | |
| Quarterly reset | | | | |

**Deliverable:** One rule to prevent re-litigation between meetings.

## Exercise 5 — Kill vanity, keep value (10 minutes)

Rewrite three vanity metrics as decision metrics.
- % tasks done → ?
- Meetings held → ?
- Story points burned → ?

**Deliverable**: New metric + threshold + decision it should trigger.

# Chapter 9 – Metrics That Matter to Leadership

Modern project environments generate thousands of data points, but only a fraction truly informs leadership decisions. Senior executives care about where investments, risks, and opportunities converge—not raw task counts or meeting minutes. For project managers, the challenge lies in translating operational complexity into metrics that drive strategic conversations. This chapter focuses on advanced practices: selecting indicators that reveal both current performance and future risks, aligning project-level objectives with enterprise priorities, and integrating early-warning systems that give leaders time to act rather than react.

## Advanced Metrics Practices for Modern PMs

### Leading vs. Lagging Indicators

Traditional reporting relies heavily on lagging metrics: budget variances, delivery timelines, or ROI after implementation. While these confirm results, they arrive too late to influence them. Leading indicators, on the other hand, reveal early patterns that predict whether a project will stay on track. Examples include:

- **Risk Signals**: Rising defect rates in the first development sprint often foreshadow quality issues downstream.
- **Engagement Metrics:** Declining stakeholder attendance in planning sessions may precede approval delays or scope disputes.
- **Throughput Trends**: Slowing cycle times during early phases sometimes indicate bottlenecks before deadlines approach.

Balancing both types prevents two extremes: reacting only after failure or chasing every early signal without context. A portfolio dashboard might combine scope stability (leading) with actual delivery variance (lagging) so leaders see both trajectory and final outcomes before reallocating funds or resources[29].

## Connecting Project OKRs to Enterprise KPIs

Many organizations adopt Objectives and Key Results (OKRs) to align day-to-day efforts with strategic ambitions. Yet, misalignment often arises when project metrics focus on output—features delivered, documents approved—rather than business impact.

To prevent this, project managers can map OKRs directly to enterprise KPIs through a structured chain:

| Layer | Example Objective | Key Results | Linked KPI |
|---|---|---|---|
| Corporate Level | Expand market share in APAC | +15% customer acquisition in Q4 | Regional revenue growth |
| Portfolio Level | Launch APAC e-commerce platform | 95% uptime, 5-day average fulfillment time | Customer satisfaction (NPS) |
| Project Level | Implement payment gateway integration | Transaction errors <1%, latency <2s | Conversion rate improvement |

[29] Zheng, Li, Claude Baron, Philippe Esteban, Rui Xue, Qiang Zhang, and Shanlin Yang. 2019. "Using Leading Indicators to Improve Project Performance Measurement." Journal of Systems Science and Systems Engineering 28 (5): 529–54. https://doi.org/10.1007/s11518-019-5414-z.

This hierarchy ensures a project manager doesn't just track whether a platform launched on schedule, but whether it contributed to customer acquisition or retention targets. By reviewing OKRs quarterly, leadership sees which initiatives truly move enterprise metrics rather than producing isolated operational wins[30].

## Early-Warning Metrics for Proactive Decisions

Even with aligned KPIs, executives need time to act before risks turn into losses. Early-warning metrics embed predictive elements into project dashboards so potential threats surface early enough for mitigation.

Some practical approaches include:

*Risk Exposure Index*

- Combines open issues, severity scores, and proximity to critical deadlines.
- If risk exposure climbs above a threshold, contingency budgets or additional resources trigger automatically.

*Cost Burn Velocity*

- Tracks how quickly budgets deplete relative to schedule progress.
- A project consuming 60% of funds with only 30% of milestones completed flags likely overruns.

*Stakeholder Responsiveness*

- Measures average turnaround time on approvals or feedback.

---

[30] Sparks, Rich. 2024. "OKRs: Objectives and Key Results." Atlassian. 2024. https://www.atlassian.com/agile/agile-at-scale/okr.

- Rising delays often precede missed integration or launch dates.

*Change Request Frequency*

- A sudden spike in scope changes early in execution correlates with future delivery instability.

Dashboards integrating these signals allow leadership to course-correct mid-flight—shifting talent, renegotiating scope, or escalating decisions—rather than learning about problems during post-mortems.

## Building Composite Metrics for Context

Isolated metrics can mislead. For example, a project might show 90% task completion while masking that the remaining 10% carries the highest delivery risk. Composite indices combine multiple signals into a single view:

- **Health Scores:** Weight delivery variance, risk exposure, and budget burn into red/yellow/green statuses for portfolio reviews.
- **Value Realization Index**: Merge cost-to-date, forecasted benefits, and time-to-value to rank initiatives competing for the same resources.

Such indices prevent executives from chasing raw numbers and instead focus attention on initiatives threatening enterprise objectives or ROI.

### Avoiding Metric Overload

Finally, advanced practices emphasize selectivity. Too many metrics dilute focus and bury decision-makers in noise. A useful rule limits dashboards to:

- Five to seven leading indicators predicting future performance.
- Three to five lagging indicators confirming outcomes.
- One composite health score for quick executive scans.

Clear thresholds—what triggers escalation, funding review, or leadership attention—transform dashboards from passive reports into active decision tools.

# Executive Decision Metrics vs. Vanity Metrics

When senior executives look at project dashboards, they focus on the business implications rather than operational statistics. This is where many reporting systems fall short. Project managers often present charts filled with percentages of tasks completed, the number of meetings held, or days ahead of schedule. While these might reflect activity levels, they rarely tell leadership whether the initiative is creating value, mitigating risk, or driving strategic objectives forward.

Reporting must shift from operational detail to decision-oriented metrics to make project data genuinely useful at the leadership level. This section examines what leaders actually need to see, how to filter out vanity metrics, and how to design dashboards that inform high-stakes decisions rather than simply documenting work done.

## What Leadership Really Wants

Different executives view projects through distinct lenses. A CFO focuses on financial impact—return on investment, cost variances, and payback periods. A COO evaluates operational efficiency, risk exposure, and resource utilization. A CMO, meanwhile, wants to understand customer impact—brand perception, retention improvements, or revenue growth tied to marketing initiatives.

Across these roles, some common priorities emerge:

- **ROI and Financial Metrics:** Leaders expect to see how every major project influences revenue growth, cost savings, or margin improvement. For example, rather than hearing a CRM project is "70% complete," they want to know that customer response times have improved by 25%, translating into a projected $1.2M in additional revenue this quarter.

- **Risk Exposure:** Executives also care about risk mitigation. A product launch delivered on time means little if regulatory compliance issues later trigger fines or delays. Metrics showing residual risk levels, audit readiness, or security incident trends carry more weight than simple delivery milestones.

- **Cost Efficiency:** Leaders frequently track cost-per-outcome rather than total spend. Two projects might each cost $500K, but if one delivers twice the impact per dollar spent, executives want that visibility.

- **Customer and Market Impact:** Metrics reflecting customer retention, Net Promoter Scores, or adoption rates directly tie projects to business growth. Operational metrics like defect counts only matter if they correlate with customer experience improvements.

When dashboards highlight these dimensions, leadership can decide where to accelerate funding, redirect resources, or pause initiatives for reevaluation.

## Filtering Vanity Metrics

Many traditional metrics fall into the vanity category because they measure effort, not impact. A few common examples include:

- **Percent Tasks Completed:** Finishing 90% of tasks means little if the remaining 10% drives 80% of the project value.
- **Number of Meetings Held**: High meeting counts may signal collaboration or inefficiency. However, this metric lacks direct business meaning.
- **Documentation Volume:** Page counts or report numbers rarely indicate quality or strategic alignment.

To replace these with meaningful indicators, project managers can adopt a value-focused metric design approach:

| Vanity Metric Example | Value-Oriented Replacement | Why It Matters |
|---|---|---|
| % Tasks Completed | % of Business Benefits Realized vs. Forecast | Connects delivery to expected ROI |
| Number of Change Requests | Impact of Scope Changes on Timeline/ROI | Measures consequence, not activity |
| Meeting Count | Decision Lead Time (average days to resolve key issues) | Reflects efficiency in moving initiatives forward |
| Lines of Code Written | Feature Adoption or Revenue per Feature | Ties development work to end-user or financial gain |

## Designing Decision-Ready Dashboards

Operational dashboards often overwhelm executives with data density while failing to provide clear signals for action. A decision-ready dashboard uses fewer, more meaningful metrics structured around three guiding principles:

## Link to Strategic Goals

Every metric should trace back to a corporate objective. If a project aims to improve supply chain efficiency, the dashboard might include metrics like cost-per-shipment, delivery lead time, and inventory turnover improvement percentages rather than raw shipment counts.

## Enable Trade-Off Discussions

Executives frequently make resource allocation decisions— whether to fund, pause, or expand initiatives. Dashboards must highlight opportunity costs, ROI comparisons across projects, and capacity constraints so trade-offs become explicit rather than buried in operational detail.

## Show Risk and Forecasts, Not Just History

Leaders prefer metrics that predict outcomes, not just report on the past. Forecasted ROI, projected delivery confidence, or scenario-based risk exposure charts help them act proactively.

## Key Elements of a Decision-Ready Dashboard:

| Metric Type | Example Metric | Leadership Use Case |
|---|---|---|
| Financial Performance | ROI Forecast vs. Actuals | Adjust funding across the portfolio |
| Risk Exposure | % High-Risk Items by Business Impact | Approve mitigation budgets or alter timelines |
| Strategic Alignment | KPI Contribution Index | Ensure projects map to quarterly strategic priorities |
| Operational Efficiency | Cost per Milestone Achieved | Benchmark productivity across initiatives |
| Customer Impact | Adoption Rate vs. Target | Link project outcomes to revenue and retention metrics |

A well-designed dashboard might present three to five key indicators at the executive summary level, each with the ability to drill into supporting detail if needed. This layered approach satisfies both leadership's need for clarity and analysts' need for data depth.

## Avoiding Common Pitfalls

Even when focusing on value metrics, some reporting habits weaken decision quality:

- **Overloading Metrics**: Too many indicators dilute focus. Leadership dashboards should prioritize a small set of actionable metrics over a large catalog of descriptive ones.
- **Static Reporting**: Quarterly PDFs or static slide decks limit interactivity. Modern tools enabling real-time drill-downs improve responsiveness and engagement.
- **Lack of Context:** Metrics without benchmarks or historical comparisons mislead. A 10% cost overrun might be acceptable if benefits doubled; without context, it appears negative.

Building dashboards around comparisons, trends, and thresholds rather than single-point numbers ensures leaders interpret data correctly before making trade-offs.

Finally, decision metrics gain power when integrated into governance rhythms. Monthly portfolio reviews might focus on ROI forecasts and risk indicators for active projects, while quarterly strategy sessions assess contribution to annual objectives. Escalation metrics—like risk exposure thresholds—trigger executive reviews outside regular cadences when immediate decisions are necessary.

This alignment between metrics and decision cycles ensures leadership time goes toward resolving trade-offs and steering strategy rather than deciphering operational reports.

## Visualization & Communication for Impact

Executives rarely have the time—or the patience—to sift through dense spreadsheets or 20-page status reports. They need clarity and context at a glance. The problem is that many project dashboards still focus on presenting raw numbers rather than shaping information into a story that informs decisions. This section explains how data storytelling principles, layered dashboard designs, and real-world examples transform metrics from passive reports into strategic decision-making tools.

### Data Storytelling Principles

Data storytelling combines three elements: data, visuals, and narrative. Without all three, project dashboards risk becoming either overly simplified or impenetrably complex.

- **Context Before Detail**: A revenue variance chart gains meaning only when compared against forecasts, previous quarters, or industry benchmarks. A 10% variance alone says little about whether it signals a problem or a success.
- **Comparisons and Trends**: Rather than displaying standalone metrics—such as customer acquisition cost—effective dashboards show how it has changed over time, whether it correlates with marketing spend, or how it compares to competitors.
- **Cause-and-Effect Framing:** Linking actions to outcomes helps leadership see consequences clearly. For example,

"Process automation reduced cycle times by 25%, contributing to a $400K cost saving over two quarters."

- **Visual Hierarchy:** Not every number belongs in bold font. Key indicators belong at the top, supporting details in smaller visuals, with optional drill-downs for analysts.

When applied together, these principles turn static metrics into narratives of progress, risk, and opportunity, allowing executives to see both the what and the so what.

## Layered Dashboards: From Boardroom to PMO

One common failure in project reporting is trying to meet all audiences with the same dashboard. Executives want quick insights; PMOs and analysts need detailed operational data. The solution is a layered approach:

| Layer | Audience | Content Focus | Decision Use |
|---|---|---|---|
| **Executive Summary** | Board, C-Suite | 5–7 KPIs tied to revenue, risk, customer impact | Funding decisions, strategic pivots |
| **Portfolio View** | Senior PMs, Functional Heads | ROI comparisons, capacity trade-offs, risk heatmaps | Resource allocation, timeline adjustments |
| **Operational Drill-Down** | PMO, Analysts, Team Leads | Task dependencies, burn rates, milestone-level details | Execution planning, issue resolution |

This structure avoids two extremes: dashboards so high-level they hide risks until it's too late, or so detailed that leadership loses sight of strategic alignment. Instead, each layer speaks the right language for its audience while keeping data integrity consistent across levels.

## Case Snapshots: When Metric Design Changed Decisions

### Case 1: Funding Reallocation in a Global IT Program

A technology portfolio once reported progress through task completion percentages. After adopting outcome-focused metrics—ROI forecast, cost-per-benefit delivered, and risk-adjusted value—leadership discovered two high-cost programs producing minimal strategic returns. Funding shifted toward three mid-sized initiatives with stronger ROI projections, increasing portfolio value delivery by 18% within a year[31].

### Case 2: Accelerating Decision Cycles in Manufacturing

A manufacturing firm redesigned dashboards to show decision lead time rather than the number of meetings held. This revealed that product launch approvals often stalled in review cycles, delaying market entry by months. Leadership introduced new escalation rules, cutting average approval time by 40% and improving first-to-market positioning for key products.

### Case 3: Project Termination Through Early-Warning Metrics

An enterprise software rollout integrated early-warning indicators—budget burn velocity, change request spikes, and stakeholder response rates—into executive dashboards. Midway, these metrics signaled declining benefit realization potential despite rising costs. Leadership terminated the project

---

[31] Zheng et al., "Using Leading Indicators," 529–54.

early, saving an estimated $3.5M in additional expenses while redirecting resources toward higher-value initiatives.

These examples highlight how metric design influences strategic outcomes, not just reporting aesthetics.

Metrics shape the conversation between project teams and leadership. When dashboards focus on leading indicators, strategic alignment, and early-warning signals, executives see risks and opportunities early enough to act. Replacing vanity metrics with decision-ready indicators ensures leadership debates revolve around trade-offs, funding shifts, and strategic priorities, not activity levels.

# Chapter 10 – Business Model Literacy for PMs

Project managers often deliver outcomes without fully seeing how those outcomes influence the organization's financial health or competitive position. Yet, as businesses face rapid shifts in markets and technology, PMs must evolve beyond timelines and deliverables to understand how their work fuels profitability and growth. Business model literacy bridges this gap. It enables PMs to see the connections between revenue streams, cost structures, strategy, and customer value.

By mastering these fundamentals, PMs can make decisions that meet project goals and strengthen the organization's long-term sustainability and strategic direction, positioning themselves as indispensable strategic partners.

## Cracking the Code of Business Models

Understanding business models isn't optional for today's project managers—it's essential. Without this literacy, projects risk becoming isolated activities instead of engines driving organizational success. This section unpacks the building blocks of business models, explores how costs and revenues shape decisions, and highlights why PMs need financial fluency to influence strategy and outcomes.

### What a Business Model Really Is

A business model defines how an organization creates, delivers, and captures value (Investopedia). It answers three fundamental questions:

- **Value Creation:** What products or services solve customer problems?

- **Value Delivery**: How are these offerings delivered efficiently to the right audience?
- **Value Capture**: How does the company earn revenue and manage costs while doing so?

Key components include:

- **Value Proposition**: The unique benefit customers gain from the product or service. For instance, Netflix's value proposition combines convenience, variety, and affordability in one platform.
- **Customer Segments:** Distinct groups served by the organization—each with different needs and price sensitivities.
- **Channels:** Ways through which the product reaches customers, from physical stores to digital platforms.
- **Revenue Streams**: How money flows in, whether through subscriptions, licensing, or direct sales.
- **Cost Structures:** Operating expenses like production, marketing, and technology infrastructure.

While often confused, strategy and business models differ. Strategy defines where the business competes and how it wins, whereas the business model explains how value and profit are structured. For PMs, this distinction matters. A project aligned with strategy but blind to revenue drivers or cost implications risks falling short of real business impact.

## Revenue Streams & Cost Structures Made Simple

Projects live or die by budgets, but budgets themselves exist within larger revenue and cost systems. For PMs, understanding these systems enables smarter trade-offs and resource allocation.

*Cost Structures:*

- **Fixed costs** remain constant regardless of output, e.g., office leases and core software licenses.
- **Variable costs** fluctuate with production or usage, e.g., raw materials and contractor hours.

Projects often influence both. For instance, a PM choosing between on-premise servers and cloud hosting affects capital expenditures (CapEx) versus operating expenses (OpEx), shaping long-term cost flexibility.

*Revenue Streams*: Organizations rarely rely on one income source. Common streams include (Corporate Finance Institute):

- **Transactional:** One-off sales, like a car purchase.
- **Recurring:** Subscriptions or memberships, such as SaaS products.
- **Licensing:** Allowing others to use intellectual property for a fee.
- **Freemium:** Basic offerings are free, advanced features are paid—common in mobile apps.
- **Advertising-Based**: Revenue from sponsored content or ad placements.

*Case Example:*

Consider a PM leading a mobile app upgrade. Choosing features that enhance user engagement might boost conversion rates for premium subscriptions—a recurring revenue stream—while ignoring them could stall revenue growth. Similarly, delaying launch dates could push marketing costs higher while reducing time in market, shrinking potential returns.

Every project decision, from scope changes to vendor selection, ripples through cost and revenue equations. PMs fluent in these dynamics become trusted advisors rather than passive executors.

### Business Acumen for PMs

Business acumen connects project activities with financial outcomes. Tools like ROI (Return on Investment), NPV (Net Present Value), and payback periods help PMs evaluate whether initiatives justify their costs.

- **ROI** measures project gains relative to costs. A 20% ROI means every dollar invested yields $1.20 in value.
- **NPV factors** in the time value of money, showing whether future cash flows outweigh initial investments.
- **Payback Period** calculates how long before a project "breaks even."

Why does this matter for PMs? Because executive decisions hinge on numbers. A project manager proposing an AI-powered customer support system who shows it will reduce service costs by 30% and pay for itself in 18 months speaks the boardroom language.

Moreover, as PMs advance, strategic literacy differentiates leaders from task managers. Senior roles demand understanding not only how to deliver but also why it matters financially and strategically. From managing schedules to shaping business outcomes, this shift turns PMs into strategic partners driving organizational growth.

# Mapping Value – From Strategy to Customer Journeys

Projects are often measured by time, budget, and scope, yet these metrics rarely capture the true value delivered to the business. Value creation happens when projects align seamlessly with strategy and translate into better customer experiences, competitive advantage, and financial outcomes.

For project managers aspiring to strategic roles, mastering tools like strategy maps, value chains, and customer journey mapping is essential. These frameworks connect day-to-day execution with the bigger picture, showing how individual decisions ripple across the organization and ultimately reach the customer.

## Strategy Maps & Value Chains Demystified

A strategy map visually links an organization's objectives across four perspectives—financial, customer, internal processes, and learning & growth—demonstrating how initiatives drive long-term goals (LinkedIn – Value Disciplines). For PMs, this clarity prevents projects from becoming siloed efforts. Instead, each task aligns with broader objectives, such as increasing market share or improving customer retention.

For example, a strategy map might show that launching a new mobile app (internal process) enhances user experience (customer), leading to higher subscription renewals (financial). A PM aware of this chain can prioritize features that directly impact satisfaction metrics rather than just technical complexity.

Closely related is the value chain, introduced by Michael Porter, which breaks down activities into primary (e.g., operations, marketing, sales) and support functions (e.g., HR, technology) to show how each adds value before the final product reaches the customer. For PMs, understanding this chain helps identify where delays, redundancies, or quality issues erode value.

- A bottleneck in procurement might delay product launches.

- Inefficient knowledge transfer between R&D and marketing could weaken go-to-market strategies.

PMs who spot these handoff gaps early can redesign workflows or propose automation solutions, ensuring smoother transitions and maximizing value at every stage (Strategic Management Insight).

## Customer Journey Maps for PMs

While strategy maps focus on organizational objectives, customer journey maps visualize the customer's entire experience—from first awareness to purchase and post-sale interactions. For PMs, this tool is invaluable because projects often shape touchpoints that directly affect satisfaction and loyalty (Strategic Management Insight).

A typical journey map includes:

- **Stages:** Awareness, consideration, purchase, onboarding, support, renewal.
- **Touchpoints**: Website visits, sales calls, product usage, customer service interactions.
- **Pain Points:** Delays, confusing interfaces, lack of support, inconsistent communication.

Consider a software implementation project. A delayed rollout might not only push budgets but also frustrate end-users waiting for critical features. Mapping the journey reveals where such delays hurt most—perhaps onboarding or support—helping PMs adjust timelines or add interim solutions to protect customer satisfaction metrics.

Moreover, journey maps highlight emotional peaks and valleys. A flawless installation followed by poor after-sales support creates a negative overall impression. PMs who anticipate these dips can work with cross-functional teams to design smoother transitions, like proactive training modules or automated support FAQs, enhancing long-term loyalty.

## Aligning Projects with Value Disciplines

Organizations typically compete along three value disciplines:

| Value Discipline | Focus Area | Project Implications |
|---|---|---|
| Operational Excellence | Efficiency, cost control, reliability | Streamlined processes, automation, lean project delivery |
| Product Leadership | Innovation, cutting-edge features | R&D initiatives, rapid prototyping, technology upgrades |
| Customer Intimacy | Personalization, deep customer relationships | CRM systems, data analytics, tailored user experiences |

For PMs, aligning projects with these disciplines ensures resources reinforce competitive advantage. A company pursuing operational excellence might prioritize ERP system upgrades to cut costs, while one focused on customer intimacy invests in AI-driven personalization.

*Example:*

A telecom company aiming for operational excellence launches a network optimization project. A PM who understands this goal won't just track milestones; they'll measure reduced downtime, faster response times, and cost savings—tangible outcomes proving strategic alignment.

Finally, aligning with value disciplines prevents conflicting priorities. Without this lens, teams may over-engineer features in a cost-leadership company or underinvest in innovation where product leadership drives success. PMs act as the bridge, ensuring every deliverable strengthens the chosen discipline rather than diluting it.

## PMs as Growth Catalysts

Organizations now expect PMs to be growth partners who understand not only how to deliver projects but also how to create measurable business value. Whether it's accelerating time-to-market, optimizing budgets, or elevating customer satisfaction, PMs influence outcomes that shape profitability, innovation, and competitive positioning. This section explores how project management directly drives organizational growth, the transition from operational execution to strategic influence, and the metrics that capture this expanded role.

### Project Management as a Profit Lever

Projects consume organizational resources, but well-managed projects generate far more than they cost. Timelines, budgets, and quality are not administrative concerns—they are levers that affect financial performance and market outcomes.

- **Timelines and Revenue Realization:** Faster project delivery means products and services reach customers sooner, accelerating revenue streams. For example, a tech firm launching a subscription-based platform three months earlier gains an additional quarter of recurring revenue while beating competitors to market.
- **Budget Discipline and Margin Protection**: Projects running over budget erode profit margins. PMs who anticipate risks, control scope creep, and negotiate better vendor contracts directly preserve profitability without compromising quality.

- **Quality and Long-Term Cost Avoidance**: Delivering substandard outputs may seem cheaper initially, but it often leads to expensive rework, warranty claims, or reputational damage. A healthcare PM overseeing electronic health record implementation knows that system downtime disrupts care and invites regulatory penalties and patient dissatisfaction.

*Example:*

A manufacturing company adopts Lean project management principles to reduce waste in production-line automation projects. By cutting implementation time by 20% and defects by 35%, the PM team enables faster plant rollouts and millions in savings on labor and scrap costs—showing how operational efficiency translates directly into financial value.

This shift reframes PMs as profit protectors and enablers rather than cost centers, positioning them closer to strategic decision-making.

## From Execution to Strategic Influence

Moving from tactical execution to strategic influence requires PMs to embrace a broader organizational perspective (RMCLS Blog, Invensis Learning). Instead of focusing solely on schedules and deliverables, strategic PMs ask: How does this project advance our competitive advantage? How will it transform customer experiences or open new markets?

**Key Shifts Include:**

- **From Deliverables to Outcomes**: A task-focused PM ensures a CRM system goes live on time. A strategic PM ensures it actually boosts customer retention and integrates with marketing analytics to improve sales conversion rates.

- **From Silos to Cross-Functional Partnerships:** Growth-focused PMs collaborate across IT, finance, operations, and marketing, ensuring projects deliver value across departments. They speak the language of each function—ROI with finance, CX metrics with marketing, compliance with legal—building credibility as enterprise-wide influencers.

- **From Risk Mitigation to Opportunity Realization**: Traditional PMs minimize threats; strategic PMs also identify opportunities, like integrating AI tools into customer support to cut costs, personalize experiences, and generate upsell potential.

*Championing Innovation:*

PMs close to operational realities often spot inefficiencies or emerging trends before executives do. By surfacing these insights and framing them in business terms—cost savings, revenue potential, or market differentiation—PMs become internal advocates for innovation rather than passive recipients of directives.

For instance, a PM managing data infrastructure might propose leveraging predictive analytics for supply chain optimization, highlighting potential savings and competitive lead times. This positions the PMO as a source of strategic ideas, not just project delivery.

## Metrics & KPIs for Strategic PMs

Growth-focused PMs track metrics reflecting financial, market, and customer value, aligning with C-suite priorities.

Key Strategic Metrics Include:

- **Return on Investment (ROI):** Demonstrates financial gains relative to project costs.
- **Time-to-Market:** Measures how quickly deliverables generate revenue or competitive advantage.
- **Customer Impact Metrics:** Net Promoter Score (NPS), adoption rates, churn reduction—critical for CX-driven initiatives.
- **Innovation Metrics:** Number of new capabilities launched, percentage of revenue from new products.
- **Risk-Adjusted Value:** Considers both upside potential and downside exposure when evaluating project portfolios.

Moreover, strategic PMs embrace continuous learning through data analytics. Post-project reviews move beyond lessons learned to predictive insights:

- *Which project types consistently deliver the highest ROI?*
- *Where do delays most often occur in the value chain?*
- *How do customer satisfaction scores shift after major implementations?*

Business model literacy transforms project managers from delivery specialists into strategic partners shaping profitability, customer value, and innovation. By mastering business fundamentals, mapping value across strategy and customer journeys, and embracing metrics that capture real impact, PMs align their work with organizational growth. This mindset shift positions PMs at the intersection of execution and strategy as organizations navigate competitive, fast-changing environments where every project can tilt the scales toward success or irrelevance.

# Part IV

# The Practice of Putting Strategy into Action

# Chapter 11 – Strategic Road mapping for Business Conversations

Roadmaps have evolved from static charts into dynamic storytelling tools that bridge the gap between strategy and execution. For project managers and business leaders, today's roadmaps do more than outline tasks—they visualize priorities, clarify dependencies, and communicate how initiatives drive business outcomes. This chapter explores practical tools, real-world examples, and advanced techniques for building strategic roadmaps that resonate with executives and stakeholders alike.

Through goal-based and theme-based approaches, practical templates, and storytelling methods, readers will learn how to turn roadmaps into narratives of growth and innovation. Whether it's choosing the right format or crafting visuals that align teams around shared objectives, this chapter provides the frameworks and examples to elevate road mapping from project planning to strategic influence.

## Building Roadmaps that Drive Real Impact

Strategic road mapping is no longer about plotting milestones on a timeline—it's about creating a shared vision that links initiatives to measurable business value. A well-designed roadmap answers critical leadership questions: What are we trying to achieve? Why now? How will this impact customers, revenue, and growth? This section unpacks how to design roadmaps that go beyond task management to become strategic communication tools.

## From Goals to Themes: The Foundation of Strategic Road mapping

Two dominant approaches dominate modern road mapping: goal-based and theme-based.

- **Goal-based roadmaps** focus on specific outcomes within a defined timeframe. They work well when deadlines, deliverables, and dependencies drive decision-making—think regulatory compliance projects or product launches with hard dates.
- **Theme-based roadmaps**, on the other hand, group initiatives under strategic priorities rather than rigid timelines. For example, themes like Customer Experience Enhancement or Digital Transformation allow flexibility as teams pursue initiatives aligned with long-term goals.

*When to Use Each Approach:*

- **Use goal-based roadmaps** when clarity and deadlines are paramount, such as launching a cybersecurity system by Q3 or meeting compliance requirements before new regulations take effect.
- **Opt for theme-based roadmaps** when priorities evolve rapidly or when executive stakeholders care more about strategic alignment than Gantt-chart precision—common in innovation-driven sectors like SaaS or fintech.

*Real-World Example:*

A fintech company rolling out AI-powered credit scoring faced skepticism from its board over budget and technical risks. Instead of presenting a date-driven roadmap filled with jargon, the project team used a theme-based roadmap centered on three pillars: Customer Trust, Risk Reduction, and Operational Efficiency. Each initiative—data security upgrades, algorithm audits, pilot testing—was tied to these themes rather than arbitrary deadlines. The result? Executives quickly saw why each step mattered strategically, securing funding and accelerating rollout approvals.

This example highlights that roadmaps must speak the language of business priorities, not just project phases.

## Roadmap Templates and Tools for Today's Leaders

The explosion of roadmap tools means PMs and executives no longer start with blank slides. Platforms like Aha!, Roadmunk, ProductPlan, and ProjectManager offer templates designed for different leadership needs—strategic planning, portfolio management, or customer-focused initiatives.

*Popular Formats Include:*

| Format | Best For | Strengths |
|---|---|---|
| Timeline | Fixed-date product launches, compliance initiatives | Clear sequencing, deadline accountability |
| Swimlane | Cross-functional projects with parallel workstreams | Visualizes ownership, reduces silo conflicts |
| Kanban-Style | Agile teams managing iterative releases | Flexible, easy to update dynamically |
| Outcome-Based | Strategy-driven initiatives with evolving priorities | Ties actions directly to KPIs and outcomes |

A consumer electronics firm compared two formats for its global product launch:

- **The timeline roadmap** clarified when engineering, marketing, and supply chain milestones aligned, helping operational teams meet shipping deadlines.
- **The outcome-based roadmap** highlighted strategic KPIs—market share targets, customer satisfaction goals—enabling executives to focus on business impact rather than internal complexity.

Leadership ultimately adopted a hybrid roadmap, using timelines for operational planning but framing executive discussions around outcomes like revenue growth and brand positioning. This flexibility proved essential when pandemic-related delays forced schedule adjustments without derailing strategic objectives.

**Practical Tip:** Start with one roadmap format, but don't hesitate to layer views. Operational teams and executives often need different lenses—tactical vs. strategic—of the same initiatives.

## Turning Roadmaps into Business Narratives

A roadmap stuffed with tasks and dates rarely inspires action. Executives want to understand how initiatives deliver value, not just when they happen. This is where storytelling with data transforms roadmaps from static visuals into compelling business narratives.

*Key Principles for Narrative-Driven Roadmaps:*

- **Link every initiative to business outcomes**: Instead of *"Upgrade CRM system,"* frame it as *"Enhance customer retention by 15% through CRM automation."*

- **Show value progression:** Use milestones to illustrate how each phase builds toward strategic goals—e.g., cost savings, new revenue streams, customer satisfaction gains.
- **Visual simplicity with strategic depth**: Executives scan dashboards; they don't read technical reports. Use color coding, icons, and short labels to communicate layers of complexity without overwhelming.

*Example:*

A SaaS company preparing for Series B funding presented two roadmaps:

1. A feature-heavy timeline detailing development sprints.
2. A business narrative roadmap linking features to market expansion, churn reduction, and revenue acceleration.

Investors engaged far more with the second roadmap because it told a growth story, not a technical tale.

*Practical Application:*

- Use metrics like ROI projections, customer adoption rates, or cost-avoidance estimates alongside milestones.
- Replace internal jargon with business-friendly language: "Market Expansion Phase" resonates more than "Release Sprint 4.2."
- Add checkpoints that quantify progress—e.g., "Pilot deployment completed with 95% uptime" before moving to full rollout.

By framing roadmaps as narratives, PMs elevate conversations from what we're building to why it matters.

# Roadmaps as Leadership and Communication Tools

The most powerful roadmaps shape conversations, drive alignment, and provide leaders with the visibility they need to make confident decisions. When used strategically, roadmaps transform from planning visuals into leadership platforms where strategy, execution, and accountability converge.

Let's find out how to present roadmaps effectively in executive settings, use them to align cross-functional teams, and apply them as dynamic tools for advanced leadership techniques like scenario planning and risk visualization.

## Facilitating Executive Conversations Through Roadmaps

Executives rarely want to see every detail of a roadmap. They want to understand how initiatives connect to business outcomes, financial priorities, and competitive advantage. Yet many project teams deliver roadmaps overloaded with tasks, technical labels, and timelines—leaving leadership disconnected from the strategic narrative.

*What Leaders Want vs. What Teams Often Build*

| Executives Want to See | Teams Often Build |
|---|---|
| Strategic outcomes and business impact | Task lists and feature releases |
| High-level milestones tied to KPIs | Technical deadlines without context |
| Clear priorities and trade-offs | Overcrowded timelines with equal-weighted initiatives |
| Visual clarity for quick decision-making | Dense Gantt charts with minimal storytelling |

Bridging this gap requires reframing roadmaps as decision-support tools. For example, instead of listing "Phase 2 CRM Features," label it as "Boost Customer Retention – Phase 2," supported by data on projected churn reduction or revenue growth.

*Example: Healthcare PMO Alignment*

A healthcare PMO leading a digital transformation faced constant friction between IT, finance, and operations. Quarterly roadmap reviews were restructured around strategic themes—improving patient experience, reducing operational costs, and ensuring compliance—rather than siloed project lists. Each milestone showed its timeline and the business metric it influenced, such as "10% reduction in patient wait times."

As a result, leadership conversations shifted from debating technical details to agreeing on strategic priorities. Funding approvals accelerated, and departments stopped competing for resources because the roadmap clarified how each initiative served the larger vision.

*Practical Tips for Executive Conversations:*

- **Use dashboards with drill-down options**: High-level views for executives, detailed layers for project teams.
- **Tie milestones to KPIs:** "Launch telehealth platform – Projected ROI: 18 months" tells a stronger story than "Phase 3 Deployment."
- **Keep visuals clean:** Use three to five strategic pillars rather than sprawling lists of initiatives.

## Driving Cross-Functional Accountability and Transparency

Siloed execution is one of the biggest threats to strategic success. Roadmaps eliminate this problem when used as shared accountability platforms rather than static documents buried in slide decks.

### Real-Time, Shared Digital Roadmaps

Modern tools like Aha!, Roadmunk, and ProductPlan allow multiple departments to collaborate on a single roadmap with real-time updates. These platforms ensure that when priorities shift—due to market changes, budget constraints, or new opportunities—everyone sees the impact immediately.

### Benefits of Shared Roadmaps:

- **Transparency:** Everyone understands how decisions affect timelines, budgets, and dependencies.
- **Alignment:** Marketing sees when product features launch; finance sees when costs peak; operations sees when workloads increase.
- **Accountability:** Clear ownership is assigned to each milestone, preventing tasks from "falling through the cracks."

### Case Example: Global Manufacturing Firm

A multinational manufacturer struggled with overlapping digital initiatives across regions, leading to duplicated efforts and resource conflicts. By moving to a centralized, collaborative roadmap, regional IT heads, supply chain leaders, and product managers could all see dependencies and negotiate priorities before conflicts escalated.

This shift reduced redundant projects by 25% in the first year and freed budget for innovation initiatives previously sidelined by resource constraints. Leadership praised the roadmap not just as a planning tool but as a negotiation table for cross-departmental alignment.

*Practical Tool Spotlight:*

- **Aha!:** Great for strategic planning with goal and initiative hierarchies.
- **Roadmunk:** Strong visualization capabilities for executive-ready roadmaps.
- **ProductPlan:** User-friendly interface for collaborative, real-time updates.

## Advanced Leadership Applications: From Vision to Execution

For senior leaders, roadmaps increasingly serve as dynamic strategy tools rather than static plans. They provide a platform for scenario modeling, risk analysis, and strategic trade-offs— capabilities essential in volatile markets.

*Techniques for Advanced Leadership Use:*

- **Scenario Planning**: Test best-case, worst-case, and likely scenarios directly within the roadmap. For example, how would a six-month supply chain delay affect a product launch, revenue forecasts, and marketing campaigns?
- **Risk Visualization**: Color-code or tag initiatives by risk level—financial, operational, or market-related—so leaders can see exposure at a glance.
- **Priority Trade-Offs**: Simulate what happens if funding shifts from one strategic pillar to another, helping executives make data-driven decisions about where to invest.

Before entering a new region, a telecom company used roadmap simulations to test multiple approaches: phased rollout, full launch, or partnership-led entry. Each version displayed costs, timelines, and projected market share gains. Executives compared scenarios side-by-side, selecting a phased approach that minimized risk while validating demand before full-scale investment.

The roadmap evolved into a living strategy document, updated quarterly as real-world data replaced assumptions. This iterative process reduced decision-making uncertainty and aligned executives, investors, and operational teams around a single source of truth.

*Leadership Mindset Shift:*

When leaders start using roadmaps this way, they stop asking, *"Are we on schedule?"* and start asking, *"Are we investing in the right opportunities, and how will changes affect our strategy?"*

Strategic road mapping has moved far beyond timelines and task lists. When designed as dynamic, collaborative, and narrative-driven tools, roadmaps help leaders connect vision to execution with clarity and confidence. As organizations face accelerating change, roadmaps become not just planning artifacts but engines of strategic agility, ensuring that every initiative contributes to business growth, innovation, and long-term resilience.

# Chapter 12 – Leading with Influence, Not Authority

Leadership in modern organizations rarely comes with unlimited authority. As businesses grow complex and globally dispersed, leaders increasingly face the challenge of driving results across teams they do not directly control. Traditional command-and-control models fail in environments defined by silos, competing KPIs, and diverse priorities.

This chapter explores how leaders can succeed through influence rather than authority—leveraging trust, collaboration, emotional intelligence, and strategic communication to align people toward common goals.

## Mastering Cross-Silo Leadership

Cross-silo leadership has become essential for project managers, department heads, and executives navigating today's matrixed organizations. Without it, even the most ambitious initiatives risk becoming fragmented efforts, slowed by bureaucracy, misaligned incentives, or cultural walls between teams. Mastering influence across silos requires leaders to understand why these barriers exist, develop tools to unite diverse stakeholders, and build trust that transcends organizational charts.

### The Reality of Siloed Organizations

Silos often emerge naturally as organizations scale. At first, they serve practical purposes—specialization improves efficiency, departments focus on their metrics, and expertise deepens within functional areas. Over time, however, these silos harden into structural and cultural barriers:

- **Specialization**: Marketing, IT, finance, and operations speak different "languages," leading to misunderstandings and competing priorities.
- **Competing KPIs**: Sales chases revenue growth while operations focuses on cost control; HR emphasizes engagement while finance enforces budget discipline. Each department optimizes for its own success, sometimes at the expense of enterprise goals.
- **Cultural Barriers**: Geographic dispersion and differing leadership styles create microcultures resistant to outside influence.

For project managers, the mental toll can be significant. They are expected to deliver cross-functional results without direct authority over siloed teams. This often leads to stress, burnout, and decision fatigue as they navigate conflicting priorities, unclear escalation paths, and political undercurrents[32].

Many PMs report spending more time mediating disagreements than executing project plans—a clear sign that leadership across silos demands more than technical expertise; it requires emotional intelligence, persuasion skills, and strategic communication.

## Building Coalitions Across Silos

Breaking silos does not happen through formal memos or executive mandates alone. It requires coalitions—networks of

[32] Boller, Dr Max. 2024. "The Mental Toll on Project Managers in Highly Functional, Siloed Organisations." Institute of Project Management. Institute Project Management. December 16, 2024.
https://instituteprojectmanagement.com/blog/the-mental-toll-on-project-managers-in-highly-functional-siloed-organisations/.

stakeholders aligned around shared goals rather than departmental agendas.

*Techniques for Creating Shared Goals:*

- **Co-creation Workshops:** Bring stakeholders from different departments to define success metrics collectively. When teams co-own goals, resistance decreases because outcomes feel shared rather than imposed.
- **Shared OKRs (Objectives and Key Results)**: Instead of marketing tracking leads and sales tracking revenue separately, align both on a joint objective like "Increase qualified leads-to-sales conversion by 20%."
- **Neutral Facilitation:** Use cross-functional steering committees or project PMOs to mediate conflicts and keep discussions objective rather than political.

*Vision-Setting and Storytelling*

Numbers inform, but stories inspire. Leaders who connect initiatives to a compelling narrative create emotional buy-in across departments. For example:

- **Instead of saying,** *"We're implementing new software,"* reframe it as, *"This platform will reduce customer complaints by 30%, freeing your teams from repetitive tasks to focus on innovation."*
- **Use before-and-after scenarios**: Show how life improves for each department after the change, whether it's reduced manual reporting for finance or faster customer response times for sales.

A global retailer used storytelling effectively when launching an omnichannel strategy. Rather than presenting a deck full of KPIs, leaders told the story of a frustrated customer navigating disconnected online and in-store experiences. They then painted the picture of a seamless future state, turning abstract projects into relatable human outcomes. Departments rallied behind the vision because they saw its impact beyond spreadsheets.

## Trust and Credibility as Leadership Currency

Influence across silos depends less on titles and more on trust and credibility. People follow leaders they believe in, not just leaders with authority.

### Building Psychological Safety

Teams collaborate openly when they feel safe to share concerns without fear of blame. Leaders foster this by:

- **Admitting their own mistakes**, signaling that learning is valued over perfection.
- **Encouraging dissenting opinions** during planning phases so risks surface early rather than derailing execution later.

Google's Project Aristotle famously found psychological safety to be the number one predictor of high-performing teams. Without it, cross-functional collaboration becomes superficial, with real issues buried under polite silence.

### Transparency and Consistency

Trust grows when leaders communicate honestly about constraints, trade-offs, and decision rationales. Hidden agendas breed suspicion; open communication creates alignment even when decisions are tough.

- **Transparency Example**: A project manager facing budget cuts shared openly with all departments how scope reductions would affect timelines and deliverables.

Because the process was transparent, stakeholders stayed cooperative despite setbacks.

- **Consistency Example**: Leaders who change priorities weekly lose credibility fast. Influence strengthens when decisions and behaviors remain stable over time, even under pressure.

Ultimately, trust acts as a force multiplier for influence. A leader respected for fairness and integrity needs fewer meetings, fewer approvals, and less escalation because stakeholders willingly align behind their vision.

Mastering cross-silo leadership requires balancing logic with empathy, strategy with storytelling, and process with trust. Leaders who develop these skills turn fragmented organizations into collaborative networks capable of executing complex initiatives without relying solely on positional authority.

## Influence, Power, and Organizational Politics

Navigating organizational life requires far more than technical skills or positional authority. Leaders today often find themselves managing complex webs of stakeholders—each with their own goals, resources, and personalities—without the luxury of clear hierarchies. In this environment, influence becomes the currency of leadership, and understanding power dynamics is essential for driving outcomes ethically and effectively.

### Understanding Sources of Power

Power in organizations is no longer confined to corner offices or executive titles. Research in organizational psychology identifies several distinct sources of power that leaders can leverage depending on the context:

| Source of Power | Description | Modern Application |
|---|---|---|
| **Legitimate Power** | Authority derived from position or title. | A project manager authorized to allocate budgets or assign resources. |
| **Expert Power** | Influence based on specialized knowledge or skills. | A cybersecurity lead shaping decisions during a data protection initiative. |
| **Referent Power** | Influence from personal charisma, likability, or reputation. | A respected team leader trusted across departments regardless of formal authority. |
| **Informational Power** | Control over valuable data or insights others need for decision-making. | A business analyst providing market intelligence shaping strategic priorities. |

In practice, expert, referent, and informational power often outweigh legitimate power in cross-functional settings. For instance, a project manager without direct authority over engineering can still drive decisions by offering unique customer insights (informational power) or by earning trust as a collaborative problem-solver (referent power). Leaders who rely solely on title-based authority risk resistance; those who cultivate multiple power bases gain adaptability and credibility across diverse stakeholders.

## Influence Techniques Beyond Authority

Once leaders understand power dynamics, the next challenge is turning that power into constructive influence. Research in psychology and behavioral economics highlights several techniques that work across cultures and organizational levels:

### Reciprocity

People tend to return favors. Leaders who help stakeholders achieve small wins—like sharing resources, offering expertise, or publicly recognizing contributions—build goodwill that makes others more willing to support future initiatives.

### Framing

How information is presented shapes decisions. For example, proposing a new CRM system as "reducing customer churn by 15%" creates urgency and aligns with revenue goals, while describing it as "software replacement" sounds like a cost center. Strategic framing connects proposals to stakeholders' priorities, increasing buy-in.

### Coalition-Building

Influence strengthens when multiple voices advocate the same idea. Leaders who build alliances across departments before big meetings reduce resistance because proposals appear as collective needs rather than personal agendas.

### Social Proof

People look to peers when making decisions under uncertainty. Highlighting that "three other divisions have already adopted this system with great results" creates momentum by showing alignment across the organization.

### Scarcity and Timing

Opportunities framed as limited or time-sensitive often receive faster attention. For instance, emphasizing that a government grant expires in six months can accelerate decisions on sustainability projects.

A marketing director sought funding for a customer analytics platform but faced skepticism from finance leaders. Instead of demanding approval, she reframed the proposal as a revenue opportunity by showing how similar tools increased conversion rates by 20% in two other regions (social proof + framing). She also secured informal support from the sales VP beforehand (coalition-building). By the time the proposal reached the CFO, it came with allies and data-driven urgency—earning approval without escalating conflicts.

## Managing Organizational Politics Without Losing Integrity

Politics in organizations often carries a negative reputation, associated with backroom deals or personal agendas. Yet some level of politics is inevitable when resources are limited, priorities compete, and stakeholders hold differing visions for success. The challenge is navigating this landscape ethically, preserving both influence and integrity.

*Common Political Scenarios:*

- **Hidden Agendas:** A department resists a project not because of budget but because it threatens their control over processes.
- **Stakeholder Conflicts**: Two executives want the same resources for competing initiatives.
- **Priority Shifts:** Leadership changes redirect funding halfway through major projects.

- **Transparency:** Share information openly rather than hoarding it for personal advantage. This builds trust even among stakeholders with competing interests.
- **Principled Negotiation:** Focus on shared organizational goals rather than personal wins. Frame discussions around "What benefits the enterprise?" instead of "What benefits my team?"
- **Boundary Spanning:** Build relationships beyond immediate teams to understand perspectives early, preventing surprises in decision-making forums.
- **Data-Driven Arguments:** Use evidence to depersonalize conflicts. When trade-offs are framed with metrics—ROI, risk reduction, compliance requirements—decisions feel less political and more objective.

*Example:*

 A project manager overseeing a digital transformation faced IT leaders resisting cloud adoption due to perceived loss of control. Instead of forcing the issue, the PM organized workshops where IT could define security requirements for the new system. This approach acknowledged their concerns (transparency), framed the project as enhancing rather than reducing IT's role (framing), and built a coalition across security, operations, and finance teams. The result: political resistance gave way to collaborative problem-solving without escalating tensions.

# Emotional Intelligence – The Leadership Superpower

Influence without emotional intelligence is like strategy without execution—ambitious but ineffective. Leaders working across silos, managing organizational politics, or guiding change cannot rely solely on technical expertise or authority. They need the ability to read situations, manage emotions (their own and others'), and respond in ways that build trust rather than friction. Emotional Intelligence (EI) provides this foundation. Research consistently links high EI with stronger leadership outcomes, including better decision-making, reduced turnover, and higher employee engagement (Medium; The Polyglot Group).

## The Science Behind Emotional Intelligence (EI)

Emotional Intelligence refers to the capacity to recognize, understand, and manage emotions effectively—in oneself and in relationships. Neuroscience shows that emotions influence not only interpersonal dynamics but also cognitive processes like judgment, problem-solving, and risk assessment. When leaders lack emotional regulation, stress hijacks decision-making, leading to impulsive choices or avoidance behaviors.

Conversely, emotionally intelligent leaders manage pressure without transmitting panic to their teams. They recognize when fear or resistance drives stakeholder reactions and address root concerns rather than escalating conflicts. Studies in organizational psychology demonstrate that leaders with high EI foster psychological safety, enabling open communication and collaboration even under uncertainty. This directly impacts team cohesion, as employees feel heard, respected, and motivated to contribute solutions rather than defend positions.

## EI Competencies for Influential Leaders

While EI encompasses multiple dimensions, four core competencies consistently emerge as critical for leaders seeking influence over authority:

| EI Competency | Leadership Impact |
|---|---|
| Self-Awareness | Recognizing personal triggers, biases, and emotional patterns before they derail conversations. Enables leaders to approach conflicts with clarity rather than defensiveness. |
| Empathy | Understanding others' perspectives and emotions. Empathy turns resistance into dialogue by showing stakeholders they are valued, not overridden. |
| Relationship Management | Building trust, resolving conflicts, and inspiring collaboration across diverse teams. Leaders strong in this area connect organizational goals to individual motivations. |
| Adaptability | Staying flexible amid changing priorities, market shifts, or unexpected obstacles. Adaptable leaders model resilience, encouraging teams to embrace rather than fear change. |

*Example:*

During a post-merger integration, a senior leader noticed rising tension between legacy teams. Instead of pushing deadlines harder, she conducted listening sessions to understand employees' concerns. By acknowledging emotions before discussing process changes, she built credibility and reduced turnover—achieving integration goals faster than mergers driven solely by operational mandates.

## Shaping Culture Through Emotional Intelligence

Emotionally intelligent leaders do more than manage tasks; they shape the culture where collaboration and innovation thrive. Three cultural shifts often follow leaders with strong EI:

Reduced Burnout: Leaders who sense early signs of team fatigue can redistribute workloads, secure resources, or reset timelines before stress turns into attrition.

Higher Engagement: Employees are more likely to commit discretionary effort when leaders connect work to meaningful goals and recognize contributions authentically.

Psychological Safety: Teams innovate when mistakes become learning opportunities rather than career risks. EI-driven leaders normalize constructive feedback instead of blame.

*Case Illustration:*

A technology company facing aggressive product timelines saw burnout rising among engineers. Instead of enforcing weekend work, the project leader collaborated with HR to introduce flexible schedules and peer-recognition programs. Productivity not only recovered but employee engagement scores improved, demonstrating how emotional intelligence strengthens both results and retention.

Ultimately, Emotional Intelligence transforms leadership from transactional coordination to human-centered influence. By combining self-awareness, empathy, relationship skills, and adaptability, leaders create environments where collaboration replaces resistance, innovation thrives under pressure, and people follow not because they must—but because they trust, respect, and believe in the vision being shared.

# Practical Applications, Tools & Real-World Examples

Theory alone doesn't build influence—practice does. This section turns concepts from earlier parts of the chapter into hands-on techniques, tools, and scenarios leaders can use immediately.

Whether managing cross-silo tensions, mapping power dynamics, or developing emotional intelligence, these applications help translate learning into action.

## Case Study: From Silo Conflict to Collaborative Launch

Imagine a project manager leading a new digital product launch involving IT and marketing teams—two departments often separated by culture, priorities, and metrics.

*Challenge:* IT prioritizes system stability, while marketing demands rapid feature rollouts for competitive campaigns. Meetings devolve into finger-pointing over delays.

*Approach:*

- **Stakeholder Mapping:** The PM identifies formal decision-makers (IT director, marketing VP), informal influencers (senior engineers, campaign managers), and potential allies across both teams.
- **Shared Goals Workshop**: Instead of debating timelines, the PM organizes a session to define joint success metrics—e.g., "System uptime of 99% during launch" and "Campaign readiness by Q3."
- **Storytelling:** The PM reframes the project as protecting customer experience rather than IT vs. marketing deadlines, creating emotional alignment around the brand's reputation.
- **Outcome:** With shared ownership of goals and reframed narratives, both teams collaborate on phased rollouts— IT ensures stability while marketing meets campaign timelines.

This generic scenario reflects countless real-world projects where influence, not authority, turns conflict into collaboration.

## Tool Spotlight: Influence and Emotional Intelligence in Action

*1. Stakeholder Mapping for Influence Analysis*

A simple stakeholder power-interest grid helps prioritize engagement efforts:

| Power / Influence | High Interest | Low Interest |
|---|---|---|
| **High Power** | Manage Closely (Key decision-makers) | Keep Satisfied (Sponsors, regulators) |
| **Low Power** | Keep Informed (Support teams) | Monitor (Peripheral stakeholders) |

*Application Tip:* Focus persuasion techniques—framing, coalition-building—on high-power, high-interest stakeholders first, then cascade engagement to others.

## 2. Emotional Intelligence Self-Assessment Tools

**EQ-i 2.0** or **MSCEIT** tests offer structured EI insights, but leaders can also use informal checklists:

- Do I recognize my stress triggers before reacting?
- How often do I ask for feedback on my communication style?
- Can I name three recent decisions influenced by empathy rather than pressure?

*Application Tip:* Pair self-assessments with 360-degree feedback for a fuller picture of emotional impact on teams.

## Practical Exercises for Leadership Influence

*Exercise 1: Framing for Persuasion*

- **Objective:** Practice reframing proposals to align with stakeholder priorities.
- **Scenario**: A budget request framed as cost savings vs. innovation investment.
- **Debrief**: Which version gained faster approval? Why did the framing matter?

*Exercise 2: Role-Play Stakeholder Conflicts*

- **Objective**: Build negotiation skills for competing departmental interests.
- **Scenario:** Two leaders want the same resources—one plays the project sponsor, the other the operations head.
- **Debrief**: Identify moments where empathy or coalition-building shifted the conversation from confrontation to collaboration.

*Exercise 3: Emotional Intelligence in Feedback Conversations*

- **Objective:** Strengthen empathy and self-awareness under pressure.
- **Scenario:** Delivering tough feedback to a high-performing but resistant team member.
- **Debrief**: Did the leader acknowledge emotions before discussing performance metrics? How did tone affect outcomes?

These tools and exercises transform abstract leadership advice into practical, repeatable behaviors that leaders can adapt to any organizational context.

Leading with influence rather than authority demands a blend of cross-silo collaboration, ethical power dynamics, and emotional intelligence. By practicing stakeholder mapping, mastering framing techniques, and developing empathy-driven leadership behaviors, readers can navigate complex organizations with integrity and impact. The case studies, tools, and exercises here ensure these concepts don't remain theoretical but become part of a leader's everyday toolkit—helping them build trust, align priorities, and inspire action even without formal power.

# Chapter 13 – Decision-Making in Uncertainty

Every leader faces moments when the road ahead is murky: too little data, conflicting priorities, rapidly shifting conditions. Traditional decision-making models assume clarity, but modern business rarely offers that luxury. Let's explore how leaders can diagnose decision contexts under uncertainty before rushing into action. By mapping whether they face certainty, risk, or pure ambiguity, applying risk-based thinking, tackling the "fuzzy front end," and building legitimacy across networks, leaders can match the right decision tools to the problem at hand. The goal is practical: to transform vague, high-stakes challenges into structured choices that support confident, credible decisions.

## Diagnose the Decision Context (Before You Decide)

Decision-making often fails not because leaders choose the "wrong" option but because they misdiagnose the type of decision they're facing. Managing a clear, well-defined problem requires one playbook; navigating ambiguous, politically entangled, or data-poor contexts requires another.

Before analyzing options, leaders must pause and map the decision environment—understanding whether they face certainty, measurable risks, or fundamental uncertainty, and whether organizational alignment exists to act effectively.

### Map the Terrain: Certainty, Risk, and Pure Uncertainty

The first step is to place the problem on the certainty– uncertainty spectrum:

| Decision Context | Characteristics | Decision Approach |
|---|---|---|
| Certainty | Outcomes and probabilities known; cause–effect relationships clear | Optimize efficiency: cost–benefit analysis, linear planning |
| Risk | Outcomes known, probabilities estimated (objective or subjective) | Risk-based decision tools: scenario planning, simulations |
| Uncertainty | Outcomes and probabilities unknown; ambiguity dominates | Discovery-driven planning, iterative experimentation |

- **Certainty**: For example, selecting between two vendors with fixed prices and delivery times fits here. Traditional ROI calculations work well.
- **Risk:** Launching a new product in a familiar market with historical data on adoption rates falls into this category. Probabilities guide choices even if outcomes aren't guaranteed.
- **Uncertainty:** Entering a completely new market with unknown customer behaviors represents true ambiguity—standard forecasting fails, and leaders must experiment, learn, and adapt.

The discipline of moving from uncertainty to risk involves reducing ignorance: gathering even rough probabilities, clarifying assumptions, and separating what is unknown from what is unknowable. Leaders can run premortems *("What might cause this project to fail?")* or use **Delphi techniques** to generate subjective probabilities from experts. The goal is to turn vague ambiguity into structured risk where tools and trade-offs become meaningful[33].

The Delphi technique is a structured way of gathering expert opinions when hard data is scarce. Instead of holding one big meeting where dominant voices might sway others, it uses multiple anonymous survey rounds with a facilitator summarizing responses each time. Experts revise their views in light of the group's feedback, gradually moving toward convergence on probabilities, risks, or scenarios. This method reduces bias, captures diverse perspectives, and works especially well in uncertain, high-stakes decisions where reliable forecasts don't exist.

## Risk-Based Decision Basics (the Leadership Lens)

When facing measurable risk rather than pure uncertainty, leaders can apply risk-based decision-making frameworks. The standard process typically follows five steps:

- Define the issue clearly: scope, objectives, constraints.
- Develop options rather than locking into a single plan.
- Evaluate options using risk criteria: probability, impact, cost, time, reputation, and compliance.
- Decide based on risk appetite and strategic alignment.
- Implement and monitor, updating as new information emerges.

A useful leadership tool here is the **4Ts risk response framework**:

---

[33] Hutchins, Greg. 2023. "Decision-Making under Certainty, Uncertainty, and Risk." Accendo Reliability. February 21, 2023. https://accendoreliability.com/decision-making-under-certainty-uncertainty-and-risk/.

| Response | Description | Example |
|---|---|---|
| Terminate | Avoid the risk entirely | Canceling an unprofitable product line |
| Treat | Reduce probability or impact | Adding cybersecurity controls to reduce breaches |
| Tolerate | Accept risk within appetite thresholds | Proceeding with minor cost overrun risks |
| Transfer | Shift risk to third parties | Using insurance or outsourcing high-risk tasks |

Leaders scale this formality to the significance and stakes of the decision. A small operational risk might need quick evaluation; a multimillion-dollar investment warrants scenario modeling, Monte Carlo simulations, or decision trees[34].

## When the Front End is Fuzzy

Many decisions collapse at the fuzzy front end—the chaotic early phase where problems lack definition, stakeholders disagree on objectives, and information is incomplete. Studies show that projects launched under such ambiguity often face delays, cost overruns, or failure to deliver expected value[35].

*Why it happens:*

- Unclear vision or shifting priorities
- Stakeholder misalignment on goals or success criteria

---

[34] "Risk-Based Decision Making." n.d. Risktec.
https://risktec.tuv.com/knowledge-bank/risk-based-decision-making/.

[35] "Addressing the Problem of Fuzzy Front End | PMO Advisory." n.d.
Www.pmoadvisory.com. https://www.pmoadvisory.com/blog/addressing-the-problem-of-fuzzy-front-end/.

- Innovation projects with no historical data for guidance

*De-fuzzing practices help leaders create structure before diving into solutions:*

- **Vision Anchor:** Establish a concise "north star" statement capturing the project's purpose and desired impact.
- **Change Champions:** Recruit influential supporters across departments to legitimize early decisions and build momentum.
- **Structured Discovery**: Use design thinking workshops, prototypes, or pilot studies to clarify assumptions iteratively.
- **Early Governance**: Set minimal but clear decision rights—who approves budgets, scope changes, or pivots?
- **Fit-for-Context Approaches**: Choose delivery methods (agile, hybrid, stage-gate) matching uncertainty levels rather than defaulting to rigid plans.

*Example:* A company exploring AI adoption in customer service ran a two-month pilot with a limited scope rather than committing to a full rollout. This surfaced technical risks, change management issues, and ROI assumptions early, reducing uncertainty before major investments.

## Legitimacy, Networks, and "Fiery Spirits"

Finally, decisions in large organizations unfold within socio-technical networks where formal authority competes with informal influence. Research by the Project Management Institute highlights the need to uncover "programs of action"—

the unwritten agendas, alliances, and resistances shaping decision environments[36].

Leaders should:

- **Map supporting vs. conflicting networks** around the decision.
- **Identify "fiery spirits"**—credible local champions who bridge silos, translate jargon, and rally peers.
- **Build legitimacy before demanding alignment**: small wins, transparent communication, and inclusive planning earn trust for bigger commitments later.

For example, in a global supply chain redesign, early engagement with regional managers ("fiery spirits") surfaced practical constraints that the headquarters had overlooked. Their credibility with local teams turned potential resistance into advocacy for change.

## Toolbench for Uncertain Decisions (Choose, Compare, Stress-Test)

When information is incomplete and stakes are high, leaders need decision tools that work fast, reveal trade-offs, and stay credible under scrutiny. This "toolbench" brings together practical methods—from quick option comparison tables to scenario planning and discovery-driven experiments—so decision-making under uncertainty feels structured rather than improvised.

---

[36] "Discover Conflicting Programs of Action: Fuzzy Project Management." 2025. Pmi.org. 2025. https://www.pmi.org/learning/library/conflicting-programs-action-fuzzy-projects-8535.

## Rapid Optioning & Comparative Evaluation

High-uncertainty decisions often come with too many opinions and too little clarity. The Alternatives & Trade-Off Table (sometimes called a decision or trade-off matrix) tackles this problem by turning subjective debates into visible comparisons.

*How It Works:*

- **List options**: Generate 3–5 credible alternatives rather than locking in too early.
- **Define criteria**: Cost, risk, time, customer impact, strategic fit—whatever matters most.
- **Assign weights**: Rank criteria by importance; not all factors carry equal weight.
- **Score options:** Use a simple 1–5 scale for each criterion.
- **Calculate totals:** Multiply weights by scores for a transparent, auditable result.

| Option | Cost (30%) | Risk (25%) | Time (20%) | Customer Impact (25%) | Weighted Score |
|--------|------------|------------|------------|-----------------------|----------------|
| Option A | 3 | 4 | 5 | 3 | 3.75 |
| Option B | 5 | 3 | 4 | 4 | 4.05 |
| Option C | 4 | 5 | 3 | 5 | 4.20 |

Option C might look promising here, but the real value is making trade-offs explicit so leaders can debate assumptions, not just gut feelings.

For quick decisions, a **"MCDA light"** approach—**Multi-Criteria Decision Analysis** simplified—avoids complex math. Normalize scores to a 0–1 scale, plot them visually, and run a sensitivity test: if weights shift slightly, does the preferred option change? If so, the decision is fragile and needs more discussion or data.

# Time–Cost–Quality (+Risk) Trade-Offs

The classic "iron triangle" of project management—time, cost, quality—often forces false choices. Research shows the best decisions rarely minimize one variable; they balance shifting priorities over the project life cycle[37].

- Early phases may prioritize speed to capture market windows.
- Mid-project, cost control might dominate.
- Close to delivery, quality and risk often outweigh schedule pressures.

Modern trade-off tools add risk explicitly, turning the triangle into a **TCQ+R model**:

| Factor | Possible Trade-Off | Leadership Question |
|--------|--------------------|--------------------|
| Time | Faster delivery vs. higher costs | "What's the revenue impact of a 3-month delay?" |
| Cost | Lower budget vs. potential quality or scope reductions | "What risks come with aggressive cost-cutting?" |
| Quality | Higher quality vs. longer schedules or larger budgets | "Where's the minimum acceptable quality line?" |
| Risk | Risk reduction vs. cost/time/quality trade-offs | "Which risks justify more time or budget investment?" |

Linear models assume smooth trade-offs; non-linear models show thresholds where small delays or cuts trigger big failures—a reality leaders must consider when approving schedules or budgets.

---

[37] Liberatore, Matthew J., and Bruce Pollack-Johnson. 2009. "Quality, Time, and Cost Tradeoffs in Project Management Decision Making." *PICMET '09 - 2009 Portland International Conference on Management of Engineering & Technology*, August. https://doi.org/10.1109/picmet.2009.5261996.

## Scenario Planning & Risk Shaping

When uncertainty dominates, building 3–4 plausible scenarios exposes blind spots before they cause damage:

- Baseline: Expected future if assumptions hold.
- Optimistic: Best case with favorable conditions.
- Pessimistic: Worst case with compounded risks.
- Wildcard: Low-probability but high-impact event (e.g., regulatory shock).

*For each scenario:*

- **Apply the 4Ts grid**—Terminate, Treat, Tolerate, Transfer—to clarify risk posture.
- **Define pivot thresholds** (metrics signaling when to shift strategies).
- **Set "kill criteria" to prevent sunk-cost bias**—conditions where stopping is wiser than continuing.

Adding a Value of Information (VOI) check helps leaders decide whether gathering more data is worth the cost or delay. For example, if waiting two weeks for a market survey cuts uncertainty by 50%, the VOI might justify the pause.

## Decision Quality Under Pressure

In crises, leaders often judge decisions solely by outcomes—success equals "good decision," failure equals "bad decision." But in uncertainty, process quality matters more than outcome luck.

Four levers improve decision quality under time pressure:

- **Framing**: Define the problem clearly—bad framing leads to irrelevant solutions.
- **Decision Rights**: Clarify who decides vs. who advises vs. who executes.

- **Cadence**: Use shorter, frequent decision cycles rather than one big "bet."
- **Documentation**: Record assumptions, rationale, and expected signals for later review.

This shifts postmortems from blame games to learning systems, strengthening future decisions even when outcomes disappoint.

## Discovery-Driven Planning for the Fuzzy Front End

When uncertainty dominates early stages, Discovery-Driven Planning treats decisions as learning investments rather than one-shot commitments.

*Key practices include:*

- **Minimum Viable Decision (MVD)**: Make the smallest decision that allows progress while keeping options open.
- **Staged Commitments**: Release budgets or resources in phases as confidence grows.
- **Learning Milestones**: Replace "percent complete" metrics with "assumption validated" metrics.

*Example:* Instead of funding a full product launch, a company might approve a pilot in one region. If adoption meets thresholds, funding expands; if not, the project pivots or ends early—saving resources and political capital.

# Field Guide: Worksheets, Templates & Generic Case Labs

Here are some tools to simplify uncertainty, make trade-offs explicit, and accelerate decision-making clarity.

## One-Pager Kits (Printable / Digital Templates)

- **Decision Context Card:** Map if the decision is certainty, risk, or uncertainty; add risk appetite and stakes rating for clarity before analysis.
- **4Ts Optioning Canvas:** For each option, list expected effects, residual risk, cost-benefit, and stakeholder stance to define response strategy.
- **Fuzzy Front-End Checklist:** Quick scan for vision anchor, change champions, assumptions log, discovery milestones, governance needs, and method fit.
- **Alternatives & Trade-Off Table Template:** Plug in options, criteria weights, and run sensitivity tests—see how priorities shift outcomes.
- **TCQ(+R) Trade-Off Sheet:** Slider bars for time, cost, quality, and top risks—define acceptable limits before commitments.

## Generic Case Labs (Short, Adaptable Vignettes)

| Case Scenario | Toolkit to Apply | Goal |
|---|---|---|
| Launch under policy uncertainty | Scenario planning + 4Ts grid | Stage investments by regulatory risk |
| Plant expansion trade-off | TCQ(+R) analysis + Weighted Decision Matrix | Pick make/buy path under deadline pressure |
| Fuzzy product concept (exec misalignment) | Discovery-driven planning + Legitimacy | Create alignment before scaling decisions |

| Case Scenario | Toolkit to Apply | Goal |
|---|---|---|
|  | building with local champions |  |

Each case takes 30–60 minutes for team workshops; can be scaled for larger strategy offsites.

Decision-making in uncertainty is less about finding perfect answers and more about creating structured confidence. By diagnosing decision contexts, applying risk-based tools, exploring scenarios, and embedding discovery-driven planning, leaders replace guesswork with transparency and adaptability. These field guides, one-pagers, and habit loops turn theory into action, helping organizations make smarter choices when clarity is scarce and stakes are high.

# Chapter 14 – Turnarounds, Triage, and Knowing When to Kill a Project

Projects rarely fail overnight. Warning signs emerge gradually—missed milestones, budget overruns, stakeholder frustration—until the problems become too big to ignore. This chapter provides practical tools and frameworks so leaders can assess scope, realign stakeholders, and decide whether to rescue, pivot, or shut down a struggling initiative.

## Rapid Triage & Turnaround Frameworks

When a project shows signs of trouble, you need fast, structured tools to understand what's broken and whether it's worth fixing. Here are the three practical assets that help leaders quickly evaluate health, prioritize interventions, and make informed go/no-go decisions.

### 1. Triage in 5 Steps – One-Pager Assessment

| Step | Key Questions | Output |
|------|---------------|--------|
| 1. Scope Check | Is the project delivering what was promised? | Gap analysis: planned vs. actual scope |
| 2. Risk Pulse | Are critical risks unmanaged or escalating? | Risk severity map (high/medium/low) |
| 3. Stakeholder Heat | Who's losing faith—sponsors, users, or team members? | Alignment score: green/yellow/red |
| 4. Resource Scan | Are budget, talent, or tools inadequate for recovery? | Resource sufficiency rating |

| Step | Key Questions | Output |
|---|---|---|
| 5. Time-to-Value | Will rescuing it deliver meaningful benefits soon enough? | Cost-benefit time horizon |

Think of *Triage in 5 Steps* as the ER check-up for projects: a rapid assessment before deciding on surgery or discharge. Each step takes minutes, not days.

*How to Use:* Run this checklist with the core team in a 60-minute session.

- Assign a Red/Yellow/Green status to each step.
- Green = stable, Yellow = warning, Red = critical intervention needed.

If 3 or more areas turn red, the project moves to turnaround or termination consideration immediately.

## 2. Turnaround Roadmap – Quick-Hit Actions

When triage reveals hope for recovery, leaders need a simple, staged roadmap rather than a six-month rescue plan. The following four phases deliver quick wins before scaling solutions:

*Stabilize*
- Freeze scope creep.
- Pause non-critical workstreams to stop the bleeding.
- Address immediate risks (e.g., funding gaps, team morale).

*Re-Scope*
- Revisit original objectives with sponsors.
- Drop low-value features or deliverables to focus on the essentials.
- Set new success criteria with measurable outcomes.

- Hold "truth-telling" sessions with executives and teams.
- Reconfirm budget, timelines, and accountability under the new scope.
- Secure verbal and written commitment from all key players.

*Relaunch*

- Communicate the new plan organization-wide.
- Track early milestones visibly to rebuild confidence.
- Implement bi-weekly health checks for the next 6–8 weeks.

**Tip:** Keep the entire rescue plan on one page with owners, deadlines, and metrics for transparency.

## 3. Rescue Red Flags Table – Top 10 Indicators for Intervention

Use this table as a visual dashboard during steering committee meetings.

| Red Flag | Why It Matters | Typical Signal |
|---|---|---|
| Repeated missed milestones | Delivery discipline collapsing | >2 consecutive sprints or phases delayed |
| Budget overrun >20% | Financial credibility at risk | Sponsor escalations, funding freezes |
| Stakeholder disengagement | Executive support eroding | No-shows at steering meetings |
| High team turnover | Morale + knowledge drain | Key leads resign mid-project |
| Scope creep without approvals | Governance breakdown | Constant feature additions |

| Red Flag | Why It Matters | Typical Signal |
|---|---|---|
| Risk log outdated or ignored | Issues unmanaged until too late | Risks closed with no mitigation evidence |
| Benefits no longer aligned to strategy | Project relevance in question | New priorities overshadow original goals |
| Vendor conflicts or failures | External dependencies unstable | Legal/contract disputes emerging |
| Data quality too low for decisions | Leaders flying blind | Status reports contradictory or missing |
| End-user resistance rising | Adoption risk post-delivery | Negative pilot feedback or low testing uptake |

*How to Use*: If 4 or more red flags stay red across two reporting cycles → consider pause or kill options, not just rescue plans.

# Managing Up with Candor

Projects in distress demand straight talk with executives—not sugarcoating or endless status slides. Leaders must communicate facts with solutions, showing realism without panic and confidence without denial. This section offers three practical tools: an Executive Debrief Template, a Candor vs. Optimism Framework, and Decision-Ready Briefs for tough calls.

### Executive Debrief Template – Status at a Glance

Keep updates short, visual, and solution-focused. One slide or page works best.

| Element | Content | Format |
|---|---|---|
| Project Health | Red/Yellow/Green status with 1-sentence reason | Traffic-light icon + note |
| Progress | Key milestones achieved vs. delayed | Timeline bar / checklist |
| Blockers | Top 3 issues halting progress | Bullet points, no jargon |
| Risks | Emerging risks + mitigation plans | Table: Risk / Owner / Action |
| Recovery Plan | Next 2–3 steps with dates and owners | Mini action list |

**Tip:** End with asks for executives—funding decisions, policy approvals, or risk tolerances—so meetings drive action, not just discussion.

## Candor vs. Optimism Framework

Leaders must balance honesty about problems with confidence in recovery. Too much candor sounds defeatist; too much optimism feels delusional.

| Tone | What to Include | Avoid |
|---|---|---|
| Candor | Facts, metrics, missed targets, root causes | Emotional language, blame games |
| Optimism | Recovery options, mitigation plans, early wins | Overpromising, vague "we'll fix it" claims |

*Script Example:* "We're 30% behind schedule due to vendor delays (candor). The revised plan drops non-critical features and adds a backup supplier, recovering 70% of lost time (optimism)."

Executives need clear choices, not endless analysis. Create one-slide summaries for each decision path:

- **Pivot:** Adjust scope, strategy, or resources to continue.
- **Pause:** Temporarily halt to resolve risks or re-scope.
- **Kill:** Shut down gracefully, capture lessons learned, redeploy resources.

Each slide should show:

- **Option Summary** – What it means in plain language
- **Pros/Cons** – Cost, risk, benefits in 3 bullets each
- **Recommendation** – Your call, with rationale

This ensures leadership conversations focus on decisions, not just data.

# Politically Smart Project Closures Tools, Exercises & Real-World Cues

Ending a project can be as politically sensitive as starting one. Executives worry about reputation; teams fear wasted effort. A structured, transparent closure process prevents chaos while preserving trust and turning setbacks into learning opportunities.

## Kill Criteria Checklist

Before recommending closure, use pre-agreed thresholds so decisions feel data-driven, not personal:

- **Financial:** Cost overruns exceed X% without strategic justification.
- **Strategic:** Project no longer aligns with organizational priorities.
- **Timing:** Delays push delivery beyond relevance or market window.

- **Value Realization**: Expected ROI falls below defined benchmarks.

When two or more thresholds trigger, initiate closure assessment immediately.

## Framing Guide for Tough Conversations

Shutting down a project isn't about failure—it's about strategic redirection. Here's a simple reframing tool:

| Instead of Saying | Reframe As |
|---|---|
| "This project failed." | "We're reallocating resources to higher-impact areas." |
| "We're shutting it down." | "We're accelerating organizational learning for future wins." |
| "Too many issues to continue." | "Continuing no longer delivers the intended business value." |

## Closure Playbook

Follow a three-phase closure process:

- **Decision & Communication** – Announce closure with clear rationale and next steps.
- **Knowledge Capture** – Document lessons learned, decision logs, and risk insights.
- **Team Redeployment** – Move talent to active initiatives quickly to protect morale and engagement.

### Case Snapshots

**Healthcare IT Rollout → Telehealth Pivot:** A regional healthcare provider began a large IT rollout aiming to integrate patient records across multiple clinics. Six months in, triage revealed severe scope misalignment—the system's complexity exceeded user readiness, timelines slipped, and costs ballooned. Instead of pouring more resources into a struggling project, leadership applied kill criteria: ROI projections fell below thresholds, and strategic relevance shifted with the pandemic's rise in telehealth demand.

The project was paused deliberately. Resources—budget, developers, and clinical IT staff—were quickly redeployed to launch a telehealth platform, which aligned better with urgent patient needs and delivered measurable ROI within a year.

**Consumer App Sunset → AI Initiative Redeployment**: A consumer-facing mobile app targeting lifestyle services completed pilot testing with underwhelming adoption rates and negative user feedback on features.

Rather than scaling a product with low potential, executives used decision-ready briefs to recommend sunsetting the app. The closure plan emphasized knowledge transfer—UX insights, data analytics frameworks, and technical lessons learned were captured systematically.

Project turnarounds demand speed, clarity, and courage. With triage frameworks, candid executive communication, and smart closure strategies, leaders can rescue what's viable, exit what's not, and protect both organizational value and stakeholder trust for the long term.

# Part V

# The Future: Becoming Indispensable to Leadership

# Chapter 15 – Your Strategic Growth Plan

## Self-Assessment & Competency Mapping

Before building new capabilities, you need clarity on where you stand. Strategic growth is about identifying the few skills that will make you indispensable to leadership conversations, not collecting random certifications or saying yes to every opportunity.

### Quick Competency Snapshot

Create a one-page matrix to evaluate yourself across four domains that define today's strategic project leader:

| Domain | Description | Self-Rating (1–5) | Priority |
|---|---|---|---|
| Strategic Thinking | Seeing patterns, anticipating change, connecting projects to long-term goals | | |
| Business Acumen | Understanding financial drivers, market positioning, customer value | | |
| Emotional Intelligence | Managing your emotions, reading others, building trust across silos | | |
| AI/Tech Literacy | Leveraging emerging tools and data insights for strategic advantage | | |

This table forces brutal honesty. Instead of rating everything a 3, highlight strengths that differentiate you and weak spots that could stall advancement. Invite feedback from a trusted colleague or mentor—often, they see blind spots you overlook.

## Gap-to-Growth Transfer

The next step is not to tackle every weakness. Focus on 1–2 areas where growth will generate the highest strategic return. For example:

- If you score high in technical delivery but low in business acumen, prioritize financial fluency. Understanding revenue streams and ROI can immediately elevate your voice in executive meetings.
- If emotional intelligence is your lowest score, build influence through active listening and empathetic communication. Technical skills might open doors, but EQ keeps you in the room when discussing strategy.
- If AI/tech literacy is your weakest area, commit to learning one relevant tool or framework that directly relates to your role, such as predictive analytics, workflow automation, or AI-powered project dashboards.

Map each gap to a targeted development activity:

- Business acumen gap → Partner with finance to review monthly reports.
- Emotional intelligence gap → Journal after difficult conversations and ask peers for feedback.
- AI literacy gap → Take a short course, then apply the tool in a live project.

The key is not to overcommit. One visible improvement in 90 days often has more impact than half-hearted attempts at five. Converting vague weaknesses into specific growth commitments establishes credibility as a professional actively shaping their own trajectory.

# 90-Day Business Fluency Sprint

Growth without a deadline is just aspiration. A 90-day sprint gives structure and urgency to your development, ensuring progress is visible and actionable. Think of it as running three focused "mini-projects" on yourself.

## Week-by-Week Sprint Plan

*Month 1: Learn Core Business Metrics and Stakeholder Goals*

- Shadow finance or strategy teams during their monthly reporting cycle.
- Create a cheat sheet of your organization's key metrics: revenue drivers, cost ratios, customer churn, and pipeline growth.
- In weekly reflections, ask: "How does my current project influence at least one of these metrics?"

*Month 2: Shadow/Exchange with a Senior Peer in Strategy*

- Identify one executive or senior manager whose role sits closer to corporate strategy.
- Arrange shadowing sessions, or exchange updates on how they evaluate opportunities and manage risks.
- Keep a running "insight log" of patterns in how they frame problems—especially their use of business language versus technical detail.

*Month 3: Pitch a Value-Add Initiative Using Newfound Fluency*

- Synthesize what you've learned into a small but tangible proposal: process improvement, cost-saving pilot, or customer experience upgrade.

- Present it in terms of business value, not project mechanics—focusing on impact, ROI, and strategic alignment.
- Even if it's not adopted, the act of pitching signals your shift from executor to strategic partner.

## Micro-Milestones Tracker

To avoid drift, track weekly wins:

- Week 1: Identified three core business metrics.
- Week 4: Linked current project outcomes to revenue goals.
- Week 6: Completed two shadowing sessions with a senior peer.
- Week 9: Drafted one-page initiative pitch.
- Week 12: Presented pitch to leadership.

This log becomes evidence of growth you can point to in reviews, networking conversations, or mentorship discussions.

# Mentorship, Shadowing, and Strategic Networking

Strategic growth accelerates when you deliberately learn from people ahead of you and invest in relationships that expand your reach. Tools like a mentorship canvas, shadowing checklists, and network maps turn these activities from abstract advice into practical actions.

## Mentorship Strategy Canvas

Mentorship works best when it's intentional. Identify 2–3 mentors who each play a distinct role:

- **Psychosocial Support**: Someone who helps you navigate stress, setbacks, and confidence dips.
- **Career Guidance:** A mentor who gives feedback on role choices, skill gaps, and promotion paths.
- **Modeling Influence**: A leader whose presence, communication, and decision-making you want to emulate.

Sketch a one-page canvas listing what you seek from each and what value you can offer in return—mentorship should be reciprocal.

## Executive Shadowing Checklist

Shadowing executives is more than observing calendars. Create a checklist of behaviors and signals to capture:

- *How do they frame decisions—through numbers, stories, or analogies?*
- *What questions do they ask before committing resources?*
- *How do they handle conflict or pushback in meetings?*
- *Which stakeholders do they prioritize, and how is their message tailored to each?*

Document insights in real time; over weeks, you'll see patterns that reveal how leadership influence is built daily.

## Network Mapping Prompt

Strategic careers rely on networks that open doors before you need them. Build a simple map:

- **Column 1:** Names of 10–15 influential contacts inside and outside your organization.

- **Column 2:** Their role in shaping opportunities (sponsor, gatekeeper, thought leader, connector).
- **Column 3:** One focus action per relationship this year (e.g., "Request 15-min chat about business trends," "Offer support on their upcoming initiative").

Update quarterly. Over time, this prevents your network from being accidental—it becomes a designed growth asset.

A strategic growth plan is less about lofty visions and more about practical moves executed consistently: assess your skills honestly, sprint toward business fluency, and leverage mentorship, shadowing, and networks to multiply your influence. These deliberate choices shift you from a competent manager to an indispensable partner in leadership.

# Chapter 16 – Branding Yourself as a Strategic PM

## Craft Your Strategic Narrative

Branding yourself starts with a clear narrative—not a list of tasks, but a story about the strategic value you bring. This begins with clarifying your purpose, selecting where you want authority, and learning to articulate impact in ways leadership understands.

### Purpose & Values Statement

Your "why" shapes how others perceive you. Instead of saying, "I manage projects," frame it as:

- *"I help organizations translate strategy into execution while reducing wasted effort."*
- *"I bridge business priorities with delivery teams, ensuring investments generate measurable returns."*

Document three elements: your mission (why you lead projects), your values (integrity, adaptability, transparency), and your perspective (how your work ties to business strategy). This becomes the foundation for resumes, LinkedIn headlines, and even performance reviews.

### Positions of Authority

Strategic branding requires focus. Identify one or two domains where you blend technical and business strengths. Examples:

- Digital transformation PM who connects emerging technology adoption with revenue growth.
- Change management PM who ensures strategic initiatives stick by building cultural alignment.
- Sustainability-focused PM linking projects to ESG targets.

This niche becomes your signature. Mention it consistently in your online profile summaries, "about" sections, and networking introductions. It signals depth, not just breadth.

## Storytelling Templates

Executives care about impact, not process. Instead of "I delivered the project on schedule," frame outcomes in the Challenge → Decision → Business Impact format:

- **Challenge:** *"Customer churn increased as legacy systems slowed onboarding."*
- **Decision**: *"I led the CRM replacement, aligning IT and sales on a single platform."*
- **Business Impact:** *"Result: reduced onboarding time by 30% and saved $1.2M annually."*

Use this structure in presentations, LinkedIn posts, and performance reviews. Over time, your brand shifts from executor to business thinker who drives measurable outcomes.

# Consistent, Thoughtful Visibility

A strong narrative is only valuable if others see it. Strategic PMs don't overwhelm feeds with noise—they show up consistently with substance.

## Micro-Content Drip Strategy

Post short, insight-driven updates weekly:

- A lesson learned from a board meeting.
- A metric you linked to project outcomes.
- A reflection on handling competing priorities.

Each post should answer: "What can others apply from this?" Relevance beats frequency. One sharp insight a week outperforms daily fluff.

## Community Engagement

Visibility is participating. Answer questions on company intranets, PM communities, or LinkedIn groups. A single, practical answer can reach more people than a polished article. Over time, peers begin to see you as a go-to advisor, not just another voice.

## Benchmark Role Models

Follow PM influencers who demonstrate business framing. Notice how they:

- Translate delivery into strategic narratives.
- Share consistent, valuable perspectives.
- Build authority by teaching others.

Reverse-engineer their style to find what fits your own voice. The goal is not imitation but learning how leaders frame value.

Strategic visibility isn't about being loud—it's about being remembered as the PM who connects projects to business impact.

# Institutional Brand Anchors

Personal branding gains real power when it's embedded inside your organization's structures. Leaders notice the PMs who consistently link their work to business outcomes, communicate impact in concise ways, and leave behind systems that outlast them.

## Performance Review Amplification

Don't just list tasks completed—translate them into business value. For instance:

- *"Implemented agile pilot"* → *"Reduced delivery cycle by 25%, enabling earlier product launches."*
- *"Managed ERP migration"* → *"Improved data accuracy by 40%, cutting reporting errors and saving $500K annually."*

Each line in a review should act as a business case snapshot showing return on leadership's investment in you.

## Project Impact Elevator Pitch

Opportunities to reinforce your brand often happen informally—walking to a meeting with a sponsor or during a quick executive check-in. Be ready with a 30-second impact story:

- *"The supply chain dashboard cut decision time from three days to one hour, helping us avoid stockouts last quarter."*
- *"The product re-scope saved 18% of budget while preserving 90% of customer-facing value."*

Practicing this format ensures you're always ready to reinforce your strategic position.

## Succession Branding

Strategic PMs aren't just valued for what they do—they're valued for what they leave behind. Build succession branding by:

- **Mentoring successors**: Equip rising PMs with skills you've mastered.
- **Codifying best practices**: Document playbooks, decision logs, and turnaround frameworks that others can use.

- **Enabling leadership stability:** When leaders see your absence would create a gap, you're no longer just a PM— you're an institutional anchor.

Branding yourself as a strategic PM is about becoming inseparable from the value chain: amplifying achievements in business terms, narrating impact in concise conversations, and creating systems and successors that show leadership you're not just managing projects—you're building organizational resilience.

# Appendix

## Book Club Questions for Teams

Use these prompts to turn the book into a practical, team learning program. Suggested cadence: 6–7 sessions, 60–75 minutes each. Output per session: one artifact (canvas/slide/checklist) and one "do-by-next" action.

### Session 1 — The Strategic Partner Mindset

**Purpose:** Shift from "project delivery" to "business impact."

- Where do we still manage tasks when a strategic outcome is the real goal? Name two examples.
- Which decisions in our work deserve a decision log? Draft the fields we'll capture.
- What would change if we measured success by business KPIs (e.g., revenue, margin, retention) instead of only time/budget/scope?
- Identify one high-visibility project where we can demonstrate strategic partnering in the next 30 days. What is the visible win?
- **Deliverable:** 1-page "From Delivery to Impact" shift plan (project, business KPI, first move).

### Session 2 — Business Model Literacy (Ch. 10)

**Purpose:** Tie projects to how the organization makes money.

- In one sentence, state our business model (value prop → customer → revenue stream → major costs). Where does your project touch it?
- List our top revenue streams and cost drivers. Which two levers can PMs influence this quarter? How?

- Which value discipline (operational excellence, product leadership, customer intimacy) are we pursuing? Pick one project and align its outcomes accordingly.
- Draft one metric chain from project deliverable → customer outcome → financial impact.
- **Deliverable**: Business model napkin sketch + metric chain for one live initiative.

## Session 3 — Strategic Road mapping

**Purpose:** Make roadmaps executive-ready.

- For your portfolio, would a theme-based or goal/date-based roadmap create better alignment? Why?
- Convert an existing feature list into outcome-based items with 2–3 KPIs each.
- Set one pivot threshold and one kill criterion for a critical stream.
- What one thing will you remove from the roadmap to increase focus?
- **Deliverable:** 1-slide executive roadmap (themes/outcomes, KPIs, pivot/kill).

## Session 4 — Influence, Politics & EI

**Purpose:** Win support without authority.

- Build a power–interest grid for your project. Who are the two "move-the-needle" stakeholders and what do they care about?
- Write two sentences framing your proposal in their language (benefit, risk removed, KPI moved).
- Identify one coalition partner outside your function. What reciprocal value can you offer?

- What EI behavior will you practice this month (e.g., curiosity before advocacy, naming emotions in tense meetings)?
- **Deliverable**: Stakeholder grid + 2-line influence script per key stakeholder.

## Session 5 — Decisions in Uncertainty

**Purpose**: Choose, compare, stress-test.

- Place an active decision on the certainty–risk–uncertainty spectrum. What info would move it one step toward "risk"?
- Build a trade-off table (cost, time, risk, customer impact, strategic fit). Which option wins?
- Apply the 4Ts (Terminate/Treat/Tolerate/Transfer) to the top risk; pick one action now.
- Define your Minimum Viable Decision and the learning milestone that will validate it.
- **Deliverable**: 1-page decision packet (context, options, trade-offs, 4Ts action, MVD).

## Session 6 — Turnarounds & Closures

**Purpose:** Act fast; rescue or exit with credibility.

- Run the 5-step triage on a troubled stream. What's red, and why?
- Draft a 1-page re-scope + relaunch plan or a closure rationale (business case, knowledge capture, redeployment).
- Write a 30-second decision-ready brief (pivot/pause/kill) for executives.
- **Deliverable**: Triage snapshot + relaunch or closure brief.

## Session 7 — Strategic Growth & Brand

**Purpose**: Make the shift durable and visible.

- Complete the competency snapshot (strategy, business acumen, EI, AI/tech). Pick two growth priorities.
- Outline a 90-day fluency sprint (learn → shadow → pitch). What's the week-1 action?
- Draft your impact elevator pitch (Challenge → Decision → Business Impact).
- Name two mentors and one relationship-building action you'll take this month.
- **Deliverable**: 1-page personal growth plan + 30-sec impact pitch.

# Group Practices (Use every session)

- **Start with evidence**: Bring one artifact (metric, slide, decision log entry).
- **Decide one action**: End with a single commitment and owner.
- **Inspect & adapt**: Open next session by reporting outcomes, not effort.

# Optional Team Metrics (track monthly)

- % initiatives with outcome-based roadmaps
- decisions with documented trade-offs / 4Ts
- Time from risk identification to action
- Executive briefs that led to clear decisions
- % team members with active 90-day sprint

These prompts convert reading into shared behaviors—and shared behaviors into organizational capability.

# Recommended Reading List

***The New Strategist: Shape Your Organization and Stay Ahead of Change* (2020) by Günter Müller-Stewens**

> From theory to practice, this guide equips managers and leaders with tools, competencies, and methods to execute strategy effectively in today's fast-changing environment.

***The Infinite Game* (2019) by Simon Sinek**

> Offers an "infinite mindset" framework, arguing that long-term success comes from endurance, flexibility, and continuing the game rather than trying to win it.

***Venture Meets Mission* (2024) by Arun Gupta, Gerard George & Thomas J. Fewer**

> Explores how entrepreneurship, government, and academia can align purpose with profit to tackle global challenges at scale.

***Experiential Intelligence* (2023) by Soren Kaplan**

> Highlights how life experience contributes to leadership horsepower on par with intellect and emotional intelligence, anchoring personal and organizational breakthroughs.

***The Chief Reinvention Officer Handbook* (2020) by Nadya Zhexembayeva**

> Provides a repeatable framework for continuous reinvention across organizations, ideal for projects in disruptive environments.

www.ingramcontent.com/pod-product-compliance
Lightning Source LLC
Chambersburg PA
CBHW040921210326
41597CB00030B/5149